Gamification for Tourism

ASPECTS OF TOURISM

Series Editors: **Chris Cooper** *(Leeds Beckett University, UK),* **C. Michael Hall** *(University of Canterbury, New Zealand)* and **Dallen J. Timothy** *(Arizona State University, USA)*

Aspects of Tourism is an innovative, multifaceted series, which comprises authoritative reference handbooks on global tourism regions, research volumes, texts and monographs. It is designed to provide readers with the latest thinking on tourism worldwide and in so doing will push back the frontiers of tourism knowledge. The series also introduces a new generation of international tourism authors writing on leading edge topics.

The volumes are authoritative, readable and user-friendly, providing accessible sources for further research. Books in the series are commissioned to probe the relationship between tourism and cognate subject areas such as strategy, development, retailing, sport and environmental studies. The publisher and series editors welcome proposals from writers with projects on the above topics.

All books in this series are externally peer-reviewed.

Full details of all the books in this series and of all our other publications can be found on http://www.channelviewpublications.com, or by writing to Channel View Publications, St Nicholas House, 31-34 High Street, Bristol BS1 2AW, UK.

ASPECTS OF TOURISM: 92

Gamification for Tourism

Edited by
Feifei Xu and Dimitrios Buhalis

CHANNEL VIEW PUBLICATIONS
Bristol • Blue Ridge Summit

DOI https://doi.org/10.21832/XU8212

Library of Congress Cataloging in Publication Data
A catalog record for this book is available from the Library of Congress.
Names: Xu, Feifei - editor. | Buhalis, Dimitrios, editor.
Title: Gamification for Tourism/Edited by Feifei Xu and Dimitrios Buhalis.
Description: Bristol; Blue Ridge Summit, USA: Channel View Publications, [2021] | Series: Aspects of Tourism: 92 | Includes bibliographical references and index. | Summary: "This book examines the cutting-edge concept of gamification in tourism. The chapters offer valuable insights and examples of best practice and address key issues of game mechanism and game design principles. This will be useful for students and researchers in tourism marketing, smart tourism and tourism futures, as well as industry practitioners"-- Provided by publisher.
Identifiers: LCCN 2020053217 (print) | LCCN 2020053218 (ebook) | ISBN 9781845418229 (pbk) | ISBN 9781845418212 (hbk) | ISBN 9781845418236 (pdf) | ISBN 9781845418243 (epub) | ISBN 9781845418250 (kindle edition)
Subjects: LCSH: Tourism--Marketing. | Gamification.
Classification: LCC G155.A1 G287 2021 (print) | LCC G155.A1 (ebook) | DDC 338.4/791--dc23 LC record available at https://lccn.loc.gov/2020053217
LC ebook record available at https://lccn.loc.gov/2020053218

British Library Cataloguing in Publication Data
A catalogue entry for this book is available from the British Library.

ISBN-13: 978-1-84541-821-2 (hbk)
ISBN-13: 978-1-84541-822-9 (pbk)

Channel View Publications
UK: St Nicholas House, 31-34 High Street, Bristol, BS1 2AW, UK.
USA: NBN, Blue Ridge Summit, PA, USA.

Website: www.channelviewpublications.com
Twitter: Channel_View
Facebook: https://www.facebook.com/channelviewpublications
Blog: www.channelviewpublications.wordpress.com

The policy of Multilingual Matters/Channel View Publications is to use papers that are natural, renewable and recyclable products, made from wood grown in sustainable forests. In the manufacturing process of our books, and to further support our policy, preference is given to printers that have FSC and PEFC Chain of Custody certification. The FSC and/or PEFC logos will appear on those books where full certification has been granted to the printer concerned.

Typeset by SAN Publishing Services.

Contents

Figures and Tables

Figures

Tables

Contributors

Editors

Professor Feifei Xu, PhD, is a Professor at Southeast University, Nanjing, China. She also held academic positions at Bournemouth University, University of Lincoln and University of West London in the UK. She was also a visiting scholar at Nottingham University, UK and the University of Waterloo, Canada. Her research areas include smart and digital tourism with a particular interest on gaming and gamification, sustainable tourism and pro-environmental behaviour. She has published over 70 papers in peer reviewed academic journals, including *Tourism Management, Annals of Tourism Research* and *Journal of Sustainable Tourism*. Feifei has served as a principal investigator of grants and sponsored projects with funding from such sources as the British Academy, the Chinese National Nature Science Foundation and China's Ministry of Education. She is a fellow of Royal Geography Society and UK National Higher Education Academy, and a member of Asia Pacific Tourism Association (APTA). She sits on the Standing Committee of Chinese Association of Tourism Geography and serves as an Associate Editor of the SSCI journal *Tourism Review*.

Professor Dimitrios Buhalis is a Strategic Management and Marketing expert with specialisation in Information Communication Technology applications in the Tourism, Travel, Hospitality and Leisure industries. He is Director of the *eTourism Lab* and Deputy Director of the *International Centre for Tourism and Hospitality Research*, at *Bournemouth University Business School* in England. He is the Editor in Chief of the most established Journal in Tourism: *Tourism Review*, and the Editor in Chief of the *Encyclopedia of Tourism Management and Marketing*. Professor Buhalis has written and co-edited more than 20 books and 250 scientific articles. He is one of the most cited authors and he was recognised as a Highly Cited Researcher 2020 by Clarivate™. His research pioneers smart and ambient intelligence tourism with a particular focus on innovation, entrepreneurship and destination ecosystems management. He works closely with the international tourism industry and is supporting tourism and hospitality organisations around the world with strategic marketing innovations using cutting-edge technologies.

Authors

Ilona Carati is an activity leader for the Brecon Beacons National Park Authority. She has delivered health and wellbeing sessions, with particular focus on national park and natural environments, on a variety of projects. These have included introducing geocaching to socially disadvantaged and disengaged young people, as well as Active Ageing health walks for older people.

Evrim Çeltek is an associate professor. Evrim obtained her MSc in tourism business administration from Sakarya University (Turkey) and she obtained her PhD in tourism and hotel management from Anadolu University (Turkey). Her research focuses on tourism marketing. She has published several articles, books and book chapters on mobile marketing, e-commerce, mobile commerce, advergames, digital marketing, augmented reality, virtual reality, gamification and electronic customer relationship management.

Zuhal Cilingir Uk, PhD, is an assistant professor at the Department of Tourism Management, Ondokuz Mayis University, Turkey. She holds a PhD from Karadeniz Technical University. She has conducted research in the areas of consumption culture, gender, symbolic consumption, marketing communication, consumer ethnocentrism, country of origin and marketing–tourism interaction.

Pablo Garrido Pintado has a PhD in information science from the Complutense University of Madrid, and has written a doctoral thesis entitled 'Online travel agencies: Situation and perspectives in Spanish electronic commerce'. He is currently an assistant professor of Faculty of Commerce and Tourism at the Complutense University of Madrid. In the last 15 years, Pablo has combined teaching with his professional career in advertising and digital marketing. For several years he has worked as a freelancer in the digital marketing department of Tour Diez Travel, which is one of Spain's top 10 tourism companies. He is the author of scientific articles and books on advertising, tourism marketing and audiovisual creation. He is an active member of the research group Icono 14. He has worked as visiting professor at the University of The Hague (The Netherlands) and the San Pablo-Tucuman University (Argentina).

Yasar Gultekin, MSc, is a lecturer at the Tourism Vocational School in Bafra, Ondokuz Mayis University, Turkey. He holds a business administration master's degree from Ondokuz Mayis University. Yasar has conducted research in the areas of museum marketing, research techniques in social sciences, cyptocurrencies' financial aspects and their usage in the tourism sector, recreation parks and gamification.

Marion Joppe, PhD, is a professor in the School of Hospitality, Food and Tourism Management, University of Guelph, Canada. She specialises in destination planning, development and marketing. She has extensive private and public sector experience and continues to be involved in the tourism industry, sitting on a number of local, provincial and federal task groups, committees and boards. Previously, she was president of the Travel and Tourism Research Association International; chair of the Tourism, Recreation and Economic Opportunities Workgroup, Waterfront Regeneration Trust; senior tourism policy coordinator, research manager and acting director for the Ontario Ministry of Culture, Tourism and Recreation, Tourism Policy Branch; and secretary of the Tourism Committee, OECD, Paris, France.

Shaojin Li is an economist at AECOM Asia Company. He has worked on the feasibility study for a commercial real estate project, a theme park and a resort, as well as other hospitality assets for more than two years. Shaojin holds a master's degree in management of real estate and hospitality assets from the Chinese University of Hong Kong and a bachelor's degree in tourism management from Southeast University, Nanjing, China. His research interests include virtual reality, gamification and tourism planning.

Sara MacBride-Stewart, PhD, is a Reader in Health Medicine and Society in the School of Social Sciences, and Sustainable Places Research Institute, Cardiff University, Cardiff, UK. Her main research interests are in the environment and health and well-being, exploring place-based approaches, affect, gender and interconnections between global North–South. Sara has a particular interest in the role of designated landscapes and protected areas for human well-being. She has published in diverse areas including the social, physical, emotional and sensory dimensions of human–nature relationships in Health: An Interdisciplinary Journal, Gender Organization and Work and Public Health. She has held grants with Newton-Ungku Omar Fund and Cardiff University staff research funds.

Elina (Eleni) Michopoulou is an associate professor in business management at the University of Derby, Buxton. Her research interests include technological applications and information systems in tourism, online consumer behaviour and technology acceptance, accessible and wellness tourism. Elina is the European editor for the *International Journal of Spa and Wellness*. She has over 50 publications in a variety of outlets including books and high impact journals such as *Annals of Tourism Research*, *Current Issues in Tourism* and *Information and Management*. She regularly acts as a reviewer for a number of top tier academic journals and serves on the editorial board of 10 of them.

Elin Nilsson has a PhD in marketing from the Umeå School of Business, Economics and Statistics where she currently works as an assistant professor. Her research interests are in the areas of consumer decision-making, consumer behaviour and choice, retail, service technology, digital interaction, augmented reality, gamification and social media.

Demos Parapanos, PhD, is a Lecturer in Tourism Management at the University of Cumbria. Demos's research interests include gamified systems and applications in the tourism and hospitality industry and his field of research is focused on gamification and understanding hotel visitors' motives when using hotel gamified applications. Demos's background in the hospitality industry started in 2006 as a chef in hotels, restaurants and pubs.

Clare Parsons, MSc, is sustainable development manager at the Brecon Beacons National Park Authority. Her work area encompasses economic and social well-being, inclusion, health and well-being, and sustainable tourism. Particular areas of interest are sustainable place-making and nature-based solutions. Clare has managed many funded projects, which have brought health and well-being benefits from the national park and natural environment to many groups and communities of interest who have not previously engaged. She has also developed and managed skills programmes for young people, thus enabling traineeships that support progression and workforce sustainability.

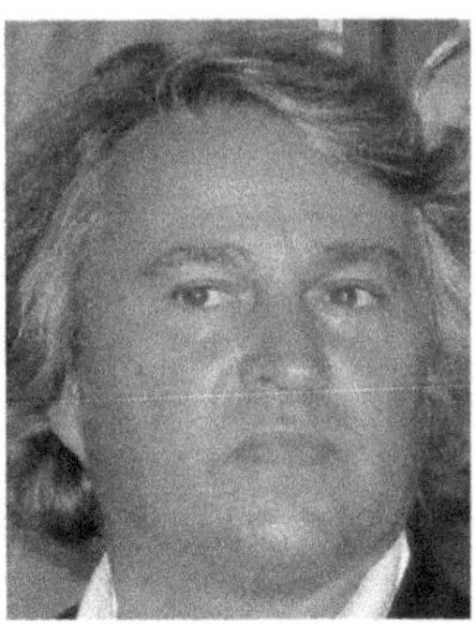

Luiz Pinto Machado first became interested in tourist activities in the hospitality sector in 1978, when he assumed leadership positions in various institutions in the industry. Luiz is a teacher at Madeira University in the fields of economics, management and tourism and at the Madeira Island Hotel and Tourism School. He is an integrated member of CEFAGE (Center for Advanced Studies in Management and Economics, Universidade de Évora), and holds degrees in hotel and tourism management and management, an MSc in strategic planning and management and tourism development, and a PhD in economics. His main research interest is tourism economics.

Ye (Sandy) Shen is an assistant professor at California Polytechnic State University, San Luis Obispo. She completed her PhD in Management at the University of Guelph, Canada and received her master's degree in urban and regional planning at Peking University. Sandy also studied hotel and tourism management at Southeast University, Nanjing,China and the Hong Kong Polytechnic University. She specialises in technology applications in tourism and hospitality, tourism experience design and experiential marketing. Sandy has several publications in top tier tourism and hospitality journals, such as *Tourism Management*, *Journal of Hospitality and Tourism Research*, *International Journal of Contemporary Hospitality Management* and *Journal of Destination Marketing and Management*.

Marianna Sigala is a professor at the University of South Australia and director of the Centre for Tourism & Leisure Management, at Adelaide, South Australia. She is an international authority in the field of technological advances and applications in tourism with numerous awarded publications, research projects and keynote presentations at international conferences. She is currently the co-editor of the *Journal of Service Theory & Practice*, the editor-in-chief of the *Journal of Hospitality & Tourism Management* and the advisor editor of the *Journal of Hospitality & Tourism Cases*. In 2016, she was awarded the prestigious The European Council on Hotel, Restaurant & Institutional Education (EuroCHRIE) Presidents' Award for her lifetime contributions and achievements to tourism and hospitality education.

Russell B. Williams is an associate professor in the College of Communication and Media Sciences at Zayed University in Abu Dhabi, UAE. His interest in game research embraces tourism and the psychology of interactive leisure experiences. He also studies product placements in videogames, destination image and the game playing tourist. Dr Williams has been teaching tourism marketing for six years and digital media design for more than 25 years. Dr Williams has held university positions in Hong Kong and the USA and holds a US utility patent for the design and gameplay of Urban Excursion: New York City.

1 Introduction: Gamification in Tourism – The Cutting-Edge Trend in Tourism

Feifei Xu and Dimitrios Buhalis

Information and communications technology (ICT) is revolutionising and disrupting tourism (Buhalis *et al.*, 2019). Technology brings a range of new disruptions in tourism and also introduces opportunities to enhance travellers' experience (Buhalis, 2020). By introducing capabilities to cocreate context dependent experiences, where location and time are critical, it is bringing a range of unprecedented interactivity opportunities (Buhalis & Sinarta, 2019; Buhalis & Foerste, 2015). Three strategically important innovations in particular have disruptive implications for service industries; namely, smart environments, cybersecurity and gamification (e.g. Bellovin, 2018; Harwood & Garry, 2015; Lu *et al.*, 2018). Gamification has become a focus of attention in an increasing number of fields, including business, education and health care. Gamification uses gaming mechanisms in a non-traditional gaming context and can have a significant impact, generating brand awareness and encouraging consumer engagement, which makes it popular in many businesses, including tourism. Gamification brings a wide range of applications and support functions with great potential for the tourism industry. Gamification can contribute to rewarding interactions and higher levels of satisfaction, leading to increased brand awareness and loyalty for tourism destinations and tourism and hospitality organisations (Xu *et al.*, 2017). Gamification is a pervasive movement, which engages consumers around the world through interactive experiences (Breidbach *et al.*, 2014).

The popularity of videogames in the past decade has made gaming attractive to a broader group of players, male and female, young and old. Empowered by the rapid development of smart mobile devices, gaming allows mobile gaming experiences and vibrant on-site communications (Gentres *et al.*, 2010), while Daisyme (2017) suggests gamification can shape everyday life by providing a fun and engaging experience for users.

Persuasive Technology and Serious Games

Technology enhances tourism experiences through the concept of 'co-creation' (Fan *et al.*, 2019; Neuhofer *et al.*, 2014). Persuasive computing technology has become popular in our daily lives (Bogost, 2007; Shuib *et al.*, 2015). Recently, serious games and the application of game elements have drawn attention from different fields, including education, health and tourism, so that they have been adopted in non-gaming contexts (Shen & Joppe, 2019; Xu *et al.*, 2014).

Serious games, as a type of persuasive technology, are computer/videogames with a set of cognitive design properties. Instead of merely providing the entertainment function of traditional games, they focus on changing user behaviour and transferring knowledge (Ghanbari *et al.*, 2015). Serious games have been widely used in training, medical applications and leisure activities in the past decade (Oinas-Kukkonen & Harjumaa, 2008). Xu *et al.* (2015: 2) suggest serious games could collect information about consumers, determining behaviour patterns, thought processes, priorities and interests. Serious games usually use gaming technologies and methodologies to engage users at a deeper level and to improve their experiences and performance. After reviewing 143 papers on games and serious games, Boyle *et al.* (2016) claim that serious games often relate to knowledge and skill acquisition, and lead to better performance.

Games, Gamification and Serious Games: A Conceptualisation

Games

Avedon and Sutton-Smith (1971) concluded that game features include voluntary participation, certain rules, different parties, potential conflicts and usually generate unequal results. Although different definitions have been used, researchers agree that a set of game characteristics can be used to define games, such as voluntary play, structure, rules, uncertain outcomes, conflict, representation, resolution and so on (Juul, 2003; Salen & Zimmerman, 2004; Seaborn & Fels, 2015: 16). Rules build up the boundaries and set up the instructions for game play (Huizinga, 2000). Conflict includes both competition and cooperation between different parties and the game system itself (Xu *et al.*, 2014). The uncertain outcome refers to winning, losing or scoring against the game's goals or other players (Crawford, 2011; Reiser, 2012). Some games include an element of luck as a random value determining the next movement of the process, while other games are driven by preference, knowledge, skill or collaboration between different players. Gaming can be very addictive since players are motivated to reach higher goals, to score points against each other and gain either material or non-material gains, such as inclusion to a hall of honour (McGonigal, 2011).

Gamification

Although there is no universally accepted definition of gamification, Deterding *et al.*'s (2011) definition, which refers to gamification as contextualising game design outside its original domain, is the most widely accepted. Zichermann and Cunningham (2011) suggest that gamification establishes a brand, engages users and influences their behaviour or knowledge by using game mechanics in areas other than the traditional gaming context. Deterding *et al.* (2011: 10) suggest that 'gamification involves applying elements of gamefulness, gameful interaction, and gameful design with a specific intention in mind'. For example, designing a treasure hunt in a tourism destination can help visitors to explore various areas and to collect points, photos, memories and experiences; however, there may not be winners or losers, as such.

Games, gamification and serious games: How do they differ?

Gamification differs from games in that the *purpose of play* is different. Gamification focuses on changing players' behaviour, engagement with their environment and co-players, who may also be fellow customers or service providers, towards achieving meaningful interactions and engagement and potentially achieving rewards. In contrast, games focus more on entertainment and pleasure (Seaborn & Fels, 2015). Alternative terms for gamification have also been used, such as 'game-based learning', 'serious games' or 'pervasive games' (Bogost, 2007; Kapp, 2012). This can further make it difficult to differentiate between the concept of 'games' and 'gamification'.

Serious games often focus on changing behaviour and influencing the user. Researchers consider these serious games as 'alternative reality games' (McGonigal, 2011) owing to the fact that reality has been added into the fantasy world (Seaborn & Fels, 2015). The incoherent use of the terminology needs more academic clarity, particularly on how to differentiate gamified systems and fully-fledged games.

Seaborn and Fels (2015: 16) further explain that 'gamefulness refers to the lived experience, gameful interaction refers to the objects, tools and contexts and gameful design refers to the practice of crafting a gameful experience'. The end result of gamification varies, as a game may have different objectives and players may use it in different ways. Gamification is about encouraging and motivating people to take actions in a structured way and follow specific rules to achieve variable outcomes (Deterding *et al.*, 2011). Games can be more about fun and entertainment, or about learning and interactively engaging in serious experience, although they can both be considered as 'funware' (Azadegan & Riedel, 2012).

Gamification mainly refers to gaming elements used outside of a gaming context. The similarities and differences between serious games and gamification mainly include the following: They are both designed for serious purposes other than pure entertainment; for example, to educate, motivate

Figure 1.1 Games, gamification and serious games

and/or persuade users, in educational, health, tourism and other contexts. Both employ games (or substantial game elements) in an effort to educate and change patterns of experience and/or behaviour; therefore, sometimes, the two terms have been used interchangeably (Fleming *et al.*, 2017).

The differences mainly include whether to use a complete game or only use a few gaming elements, and whether to use it outside of a gaming context. Serious games are still games; therefore, they provide a full gaming experience. In contrast, gamification refers to the use of game elements in non-game contexts. Gamification may not operate as a full game experience but contains gaming elements, such as a point system, in-game rewards or engaging in quests (Fleming *et al.*, 2017).

Gamification suggests the use of game design elements/mechanisms in the non-gaming context. Serious games and gamification are both designed for serious purposes, while games are mainly for pure entertainment. Serious games are still games, since they provide full gaming experiences, while gamification emphasises the use of the gaming mechanism outside the gaming context; therefore, it may not provide the full gaming experience (see Figure 1.1).

The Gamification Trend in Tourism

The terminology 'gamification' refers to the application of a game mechanism outside its original domain (Deterding *et al.*, 2011; Epstein, 2013). Through a wide range of applications and support functions, gamification has great potential for businesses, such as engaging customers in experiential co-creation and training service providers for innovative processes and functions. Seaborn and Fels (2015: 16) state that gamification has its 'roots in marketing endeavours'; for example, loyalty cards, stamp books, competition and reward membership are all early loyalty management approaches, as customers collect points to exchange for presents and gifts. Another example is Foursquare restaurant partners who supported check-ins to redeem game players' virtual rewards, such as free cupcakes and drinks (Frey, 2012). World Travel Market (WTM) (2011) stated that gamification is spreading to the travel industry and predicts it will be a popular trend for the travel industry. Research on gamification discussing its use for marketing and the service industry is emerging rapidly (Huotari & Hamari, 2012; Lucassen & Jasen, 2014; Xu *et al.*, 2014) as well as being used to help promote sustainable behaviour. Academic research on the use of gamification in the tourism context is still limited but emerging (Correa & Kitano, 2015; Sigala, 2015; Xu *et al.*, 2014). Bulencea and Egger's (2015) book on 'gamification in tourism' is a welcome addition to the theoretical development on experience design via gamification. Researchers acknowledge the importance of gaming in enhancing tourism experiences. For example, Xu *et al.* (2017) suggest gaming can foster curiosity, and encourage exploration, socialising and fun in motivating tourists to visit the destination. Swatch and Ittermann (2017) mention that gaming can also add challenges and achievements to our experiences through leaderboards, badges, level completion, collecting points and so on. The storytelling nature of games may generate fantasy in tourism experiences as well. In addition to enhancing visitor experiences, games have also been proved useful in promoting sustainable behaviour. For example, Buettel and Brook (2016) used the augmented reality game 'Ingress ' to show how gaming technology can stimulate people's interest in nature and promote ecological restoration.

Structure of the Book

The aim of this book is to investigate how the concept of gamification can transform, and has transformed, the tourism industry. The book includes two parts.

Part 1 includes three conceptual chapters on gamification theories. Chapter 1 discusses the gamification trend in tourism, and the similarities and differences between games, gamification and serious games. Chapter 2 concentrates on the concept of games and gamification, and the differences between the two. Then, it introduces game mechanics and game thinking before moving on to different types of game players. The chapter concludes by suggesting how gaming mechanisms have been adopted in the tourism industry. Chapter 3 focuses on games and gamified experiences. It reconciles and expands the two well-known conceptualisations of games, MDA (mechanics, dynamics and aesthetics) (Hunicke *et al.*, 2004) and MAST (mechanics, aesthetics, story and technology) (Schell, 2008). It proposes a combined MAPS-AIM model, which includes three circles; content, experiences and connection. The MAPS-AIM model is an important foundation of experience design in games (Costikyan, 2002).

Chapter 4 moves the discussion to the link between gamification and tourism. It focuses on gamification practices in the tourism industry, its benefits and challenges. It discusses the concept of gamification and game mechanics as well as motivations to play while summarising gamification practices. It also highlights the benefits and challenges of applying gamification in the tourism and hospitality context from the perspectives of different stakeholders. Chapter 4 provides some good examples of gamification implications in the tourism industry.

Part 2 focuses on empirical case studies on gamification, including applications in airlines, restaurants and destinations. Chapter 5 reviews the applications of gamification in tourism with a particular focus on the hospitality and airline industries. By adopting the strengths of these applications and assessing the probable opportunities and best practice, a unified gamification model is proposed. Chapter 6 expands the field by analysing a gamification application that was developed for the restaurant industry. Pinchos (a Swedish restaurant chain) has used gamification for developing its business concept including its value proposition, operations and dining experience. The gamified Pinchos app enables customers to search, order and pay for food in a fun and entertaining way through their mobile phones. The chapter analyses the gamification design of the Pinchos app by explaining how its funware enhances customers' dining experiences, their interactions with staff and the whole operation of the restaurant. Chapter 7 uses the Austria Adventure Game as a case study to explore how gamification influences destination brand awareness and user intention to travel to the destination. The quantitative analysis results suggest that gaming experiences influence brand identify and brand awareness significantly. Chapter 8 takes us to geocaching in Madeira, Portugal. Through investigating the tourist activities of geocachers, this chapter explores how they plan round trips, and look not for only one cache but a string of caches in which they can vary the order individually. The chapter seeks to identify how the development of the gamification concept and specifically geocaching can contribute to attracting more tourists and to reinforce the destination image of the island. Chapter 9 explores the implications of videogames as an advertising tool in the tourism sector, with a particular focus on the Spanish tourism market. After reviewing 10 cases of advergames and applications for mobile phones and tablets, with intensive use of gamification, the chapter compares the objectives, target audiences and marketing solutions of each game. Chapter 10 takes us to one of the most successful location-based augmented reality (AR) games, Pokémon GO. The chapter argues that Pokémon GO is a form of serious leisure and therefore a signpost for dealing with a niche form of tourism, that is, game play tourism. Using an online questionnaire for Pokémon trainers from various locations around the world, this empirical chapter explores and compares serious players and casual players regarding their game play experiences and the impact of that play on their touristic activities and knowledge. The results identify that Pokémon GO tends to attract serious leisure and therefore functions as a signpost for reaching and engaging game-playing tourists. The success of Pokémon GO has been highly reliant on nostalgia. The results also suggest Pokémon GO players travel to play and do not play to travel. Chapter 11 introduces us to a geocaching project run by the Brecon Beacons National Park Authority in Wales, UK. The chapter explores the question of how the play element of geocaching can successfully engage young people with nature, in a national park setting. The qualitative walk-along interview and observation results suggest that gaming could be useful in connecting young people to nature in both direct and indirect ways. This empirical research suggests the playfulness and game play of geocaching have a significant role in connecting young people to nature through social interaction and engagement in nature. Finally, Chapter 12 introduces the concept of augmented reality, virtual reality and mixed reality in games. It compares the advantages and challenges of applying virtual reality and augmented reality in game marketing. It then analyses some game examples of the target group, marketing benefits and marketing applications.

Conclusion

The collection of chapters in this book broaden and deepen the academic research on gamification in tourism. The book covers issues on game design elements, game player types and their motivations, game mechanism, location-based games, augmented reality and virtual reality games. It investigates how gamification can be used by airlines, restaurants, hotels and various destinations, demonstrating the complexities inherent in gamification design and gamification for marketing. Drawing on various examples and case studies, the contributors to this book reveal the cutting edge gamification trend in tourism. This book adds to our knowledge regarding this emerging trend in tourism.

References

Aparicio, A.F., Vela, F.L.G., Sánchez, J.L.G. and Montes, J.L.I. (2012) Analysis and application of gamification. Proceedings of the 13th International Conference on Interacción Persona-Ordenador. Presented at INTERACCION'12, ACM, Elche, Spain, p. 17.

Avedon, E.M. and Sutton-Smith, B. (1971) *The Study of Games*. New York: John Wiley.

Azadegan, A. and Riedel, J.C.K.H. (2012) Serious games integration in companies: A research and application framework. Proceedings of the 2012 IEEE 12th International Conference on Advanced Learning Technologies. Presented at ICALT 2012, IEEE, Rome, Italy, 485–487.

Beard, J. and Ragheb, M.G. (1983) Measuring leisure motivation. *Journal of Leisure Research* 15 (3), 219–228.

Bellovin, S.M. (2018) Toward a national cybersecurity policy. *IEEE Security and Privacy* 16 (3), 108.

Bogost, I. (2007) *Persuasive Games: The Expressive Power of Videogames*. Cambridge, MA: MIT Press.

Boyle, E.A., Hainey, T., Connolly, T.M., Gray, G., Earp, J., Ott, M., Lim, T.M., Ninaus, M.M., Ribeiro, C. and Pereira, J. (2016) An update to the systematic literature reviw of empirical evidence of the impacts and outcomes of computer games and serious games. *Computer & Education* 94, 178–192.

Breidbach, C.F., Brodie, R. and Hollebeek, L. (2014) Beyond virtuality: From engagement platforms to engagement ecosystems. *Managing Service Quality* 24 (6), 592–611.

Buettel, J. and Brook, B.W. (2016) Engress! How technophilia can reinforce biophilia to improve ecological restoration. *Restoration Ecology* 6 (24), 843–847.

Buhalis, D. (2020) Technology in tourism – from information communication technologies to eTourism and smart tourism towards ambient intelligence tourism: A perspective article. *Tourism Review* 75 (1), 267–272. https://doi.org/10.1108/TR-06-2019-0258

Buhalis, D. and Sinarta, Y. (2019) Real-time co-creation and nowness service: Lessons from tourism and hospitality. *Journal of Travel & Tourism Marketing* 36 (5), 563–582.

Buhalis, D. and Foerste, M. (2015) SoCoMo marketing for travel and tourism: Empowering co-creation of value. *Journal of Destination Marketing & Management* 4 (3), 151–161.

Buhalis, D., Harwood, T., Bogicevic, V., Viglia, G., Beldona, S. and Hofacker, C. (2019) Technological disruptions in services: Lessons from tourism and hospitality. *Journal of Service Management* 30 (4), 484–506.

Bulencea, P. and Egger, R. (2015) *Gamification in Tourism: Designing Memorable Experiences*. Norderstedt: Books on Demand.

Correa, C. and Kitano, C. (2015) *Gamification in Tourism: Analysis of Brazil Quest Game*. The National Council for Scientific and Technological Development.

Crawford, C. (2011) *The Art of Computer Game Design* (2nd edn) https://www.digitpress.com/library/books/book_art_of_computer_game_design.pdf.

Costikyan, G. (2002) I have no words and I must design: Toward a critical vocabulary for games. In F. Mayra (ed.) *Proceedings of the Computer Games and Digital Cultures Conference* (pp. 9–34). Tampere, Finland.

Deterding, S., Dixon, D., Khaled, R. and Nacke, L.E. (2011) From game design elements to gamefulness: Defining 'gamification'. *Mindtrek 2011 Proceedings*. Tampere, Finland: ACM Press.

Epstein, Z. (2013) Enterprise gamification for employee engagement. Ithaca College. Availabe at: https://doksi.net/en/get.php?lid=29219 (accessed 22 December 2020).

Fan, D., Buhalis, D. and Lin, B. (2019) A tourist typology of online and face-to-face social contact: Destination immersion and tourism encapsulation/decapsulation. *Annals of Tourism Research* 78, 102757.

Frey, T. (2012) 28 major trends for 2012 and beyond: Part 2. *Journal of Environmental Health* 74 (9), 39–43.

Fleming,T., Bavin, L., Stasiak, K., Hermansson-Webb, E., Merry, S.N., Cheek, C., Lucassen, M., Lau, H.M., Pollmuller, B. and Hetrick, S. (2017) Serious games and gamfication for mental heath: Current status and promising directions. *Frontiers in Psychiatry* 7. doi: 10.3389/fpsyt.2016.00215.

Gentres, A., Guyot-Mbodji, A. and Demeure, I. (2010) Gaming on the move: Urban experience as a new paradigm for mobile pervasive game design. *Multimedia Systems* 16, 43–55.

Ghanbari, H., Simila, J. and Markkula, J. (2015) Utilizing online serious games to facilitate distributed requirements elicitation. *Journal of Systems and Software* 109, 32–49.

Harwood, T. and Garry, T. (2015) An investigation into gamification as a customer engagement experience environment. *Journal of Services Marketing* 29 (6/7), 533–546.

Huizinga, J. (2000) *Homo Ludens: A Study of the Play Element in Culture*. London: Routledge.

Hunicke, R., Leblanc, M. and Zubek, R. (2004) MDA: A formal approach to game design and game research. Paper presented at the Proceedings of the Challenges in Games AI Workshop, 19th National Conference of Artificial Intelligence 25–29, July. San Jose, California.

Huotari, K. and Hamari, J. (2012) Defining gamification: A service marketing perspective. Proceedings of the 16th International Academic MindTrek Conference, 3–5 October. Tampere, Finland: ACM.

Juul, J. (2003) The game, the player, the world: Looking for a heart of gameness. Paper presented at Level Up: Digital Games Research Conference Proceedings, Utrecht, Netherlands.

Kapp, K.M. (2012) *The Gamification of Learning and Instruction: Game Based Methods and Strategies for Traning and Education*. San Francisco, CA: Pfeiffer.

Lu, Y., Papagiannidis, S. and Alamanos, E. (2018) Internet of Things: A systematic review of the business literature from the user and organisational perspectives. *Technological Forecasting and Social Change* 136, 285–297.

Lucassen, G. and Jansen, S. (2014) Gamification in consumer marketing – Future or fallacy? *Procedia – Social and Behaviour Science* 148, 194–202.

McGonigal, J. (2011) *Reality is Broken: Why Games Make Us Better and How They Can Change the World*. New York: Vintage.

Neuhofer, B., Buhalis, D. and Ladkin, A. (2014) A typology of technology enhanced experiences. *International Journal of Tourism Research* 16, 340–350.

Oinas-Kukkonen, H. and Harjumaa, M. (2008) Towards deeper understanding of persuasion in software and information systems. Proceedings of the First International Conference on Advances in Human-Computer Interaction, ACHI, 2008, 200–205.

Reiser, R.A. (2012) Gaming. In N.M. Seel (ed.) *Encyclopedia of the Sciences of Learning* (pp. 1335–1340). Springer.

Ritchie, J.R.B. and Hudson, S. (2009) Understanding and meeting the challenges of consumer/tourist experience research. *International Journal of Tourism Research* 11 (2), 111–126.

Salen, K. and Zimmerman, E. (2004) *Rules of Play: Game Design Fundamentals*. Cambridge, MA: MIT Press.

Schell, J. (2008) *The Art of Game Design: A Book of Lenses* (Vol. 1). Burlington, MA: Morgan Kaufmann Publishers/Elsevier.

Seaborn, K. and Fels, D.I. (2015) Gamification in thoery and action: A survey. *International Journal of Human-Computer Studies* 74, 14–31.

Shen, Y. and Joppe, M. (2018) Gamification in tourism advertising: Game mechanics and practices. *e-Review of Tourism Research*. Available at: https://journals.tdl.org/ertr/index.php/ertr/article/view/124 (accessed 24 November 2020).

Shuib, L., Shamshirband, S. and Ismail, M.H. (2015) A review of mobile pervasive learning: Applications and issues. *Computers in Human Behavior* 46, 239–244.

Sigala, M. (2015) Gamification for crowdsourcing marketing practices: Applications and benefits in tourism. In F. Garrigos-Simon, I. Gil-Pechuán and S. Estelles-Miguel (eds) *Advances in Crowdsourcing* (pp. 129–145). Springer, Cham. https://doi.org/10.1007/978-3-319-18341-1_11

WTM (2011) World Travel Market Global Trends Report. Available at: http://www.euromonitor.com/medialibrary/PDF/WTMGlobalTrendsReport2011.pdf (accessed 12 January, 2014). No longer available.

Xu, F., Tian, F., Buhalis, D., Weber, J. and Zhang, H. (2015) Tourists as mobile gamers, the gamification for tourism marketing. *Journal of Travel and Tourism Marketing* 33 (8), 1124–1142. doi: 10.1080/10548408.2015.1093999.

Xu, F., Buhalis, D. and Weber, J. (2017) Serious games and the gamification of tourism. *Tourism Management* 60, 244–256.

Zichermann, G. (2011) Intrinsic and extrinsic motivation in gamification. Gamification Co. Available at: http://www.gamification.co/2011/10/27/intrinsic-and-extrinsic-motivation-in-gamification/S (accessed 10 October 2014).

Zichermann, G. and Cunningham, C. (2011) *Gamification by Design: Implementing Game Mechanics in Web and Mobile Apps*. Newton, MA: O'Reilly Media.

Part 1

Gamification Theories

2 Gamification, Game Mechanics, Game Thinking and Players' Profile and Life Cycle

Demos Parapanos and Eleni Michopoulou

Introduction

Gamification has been identified as a major upcoming trend and is a concept that has been receiving a lot of attention recently (Xu *et al.*, 2014). The term gamification first appeared in 2008 and was introduced by Brett Terill as the use of game mechanics to other web properties to increase engagement (Deterding *et al.*, 2011; Huotari & Hamari, 2012) and has gained popularity since 2010 (Huotari & Hamari, 2012). As the term gamification is fairly recent, its definition and conceptualisation are still ongoing. Its conceptual underpinnings are often found within a marketing context (Dickey, 2005; Stapleton, 2004), although there is evidence to demonstrate that there is still confusion with similar terms such as games, achievements and rewards, and gamification. It is important, therefore, to examine the various definitional approaches to gamification and attempt to unpack its inherent complexities, in order to better understand the phenomenon.

There is often confusion between the terms games and gamification (Post, 2014). While the primary purpose of games is to entertain, gamification seeks to motivate people to change behaviours, develop new skills or engage in innovation (Post, 2014). Videogames fundamentally present a continuous process of learning to users, as they are constantly evolving and progressing their knowledge and skills (Zichermann & Cunningham, 2011). Kim (2011) highlights the importance of learning in gamified systems, and argues that learning, fun and mastery for a game are fundamental. However, these important characteristics are not always applied to gamified systems as users do not always learn something new or different (neither a cognitive nor a perceptual skill).

The most popular definition of gamification is the use of game design elements in a non-game context (Deterding *et al.*, 2011; Swan, 2012; Van

Grove, 2011). This definition seemed to be generic and a more detailed definition was presented later on stating that 'gamification is using game-based mechanics aesthetics and game thinking to engage people, motivate action, promote learning and solve problems' (Kirsh, 2014: 63). Similarly, Maedche *et al.* (2012: 186), describe gamification as the use of game design elements and mechanisms in non-game contexts to ' create a sense of playfulness... so that the participation becomes enjoyable and desirable'. Hence, gamification does not only refer to the phenomenon of applying a game mechanic to a non-game context, but it also includes a much more complex process of understanding human behaviour in order to encourage activities such as motivation and problem solving. Considering these later definitions, there seems to be an inherent and ongoing discussion between two schools of thought. Those who focus on game mechanics (such as points, badges and incentives) and those who focus on internal motivation such as game thinking and motivational design (Marczewski, 2013). This dichotomy has also been underlined by Zichermann and Linder (2010) who discuss gamification as the process of using game thinking *and* mechanics to engage audiences and solve problems.

Therefore, to be able to understand the concept of gamification more holistically, the following section will delve further into the discussion about game mechanics and game thinking.

Gamification and Game Mechanics

A common implementation of gamification is to take game mechanics, such as the scoring elements of videogames such as points, levels and achievements, and apply them to a non-gaming context such as work or education (Deterding *et al.*, 2011). The concept of gamification – using game mechanics such as rewarding users with points and badges in non-game settings – has become a core strategy for countless businesses (O'Brien, 2014). There have been claims that gamification has been defined restrictively, focusing only on the motivational power of competitiveness and achievement (Warmelink, 2014). This includes the introduction of rewards, challenges and contests, leading many gamification enthusiasts to introduce scoring systems, badges and leaderboards either among customers (marketing efforts) or employees (human resources management), showing limited understanding of common characteristics of games (Warmelink, 2014). Games try to assure and increase those feelings with reward systems that give immediate recognition to players' success, awarding them points, trophies or items on task completion. When players fail, they are expected to feel anxiety, which to some degree is acceptable, but not desirable when it transforms into frustration (Dominguez *et al.*, 2013). Gamification can be used to encourage behaviour through badge and reward systems (an aspect of gamification also known as 'pointsification') (Hanus & Fox, 2015). The term 'pointsification' has been suggested as a label for gamification systems that add nothing more than a scoring system to a non-game activity. Strohmeyer (2013) warns against confusing gamification with 'pointsification' and suggests that an individual is not going to do something that he or she is not willing to do by giving them points and badges; rather, it is important to think of gamification as a way of amplifying an existing signal. It does not make the individual do anything that were not already doing nor did not want to do, but it has to be a part of something that they already have an underlying, intrinsic interest in doing (Strohmeyer, 2013).

However, over the past decade or so, tech-savvy companies have begun exploring, adopting and refining the principles of game mechanics in order to increase sophisticated ways and get better performance from their employees, but also to encourage desired behaviour from their customers (Strohmeyer, 2013). There are numerous game mechanics that a creative designer could incorporate into a gamified system (Kirsh, 2014), especially since technology has catapulted games beyond the console and into the masses playing games such as Angry Birds, Farmville, Pokémon or Words with Friends (Al-Zaidi, 2012). The five most commonly used game mechanics are: (1) points (frequently used to award achievement and measure user's achievement in relation to others for rewarding reasons); (2) badges (ranging from easy – i.e. newbie badges are awarded to users on their first check-in – to nearly-impossible to unlock – i.e. it takes 10 movie theatre check-ins to earn the Zoetrope badge); (3) levels (i.e. encourage mobile users to level up and get better discounts for becoming more loyal patrons); (4) leaderboards (motivate and encourage users to become players); and (5) challenges (ranging from the simple to complex and often involve communal activity or group play) (Van Grove, 2011; Zichermann & Cunningham, 2011).

Game mechanics are not a panacea in themselves. Strohmeyer (2013) suggests that organisations interested in applying game mechanics to their business environment should focus on the following five principles: (1) have a measurable goal (e.g. if the purpose is to increase the number of product reviews on your website, reward users for writing product reviews by giving them points every time they do it); (2) focus on things people already want to do (reward a behaviour that is already happening); (3) measure the change (track the desired behaviour before and after gamifying it, so to measure whether the gamification is working); (4) reward incremental progress (because when individuals are rewarded for making incremental progress towards larger goals, they are encouraged to keep going); and (5) make it social by giving people the opportunity to share their accomplishments and badges, which adds meaning and significance to their achievements. Examples of gamification mechanics applications in other industries include the Nike+ platform that engages fitness enthusiasts to take their workouts to the next level, the Khan Academy that uses gamification to enhance the learning experience for students and Codeacademy that uses gamification to crowd source ideas from inventors for product development (Buckner, 2014).

Gamification and Game Thinking

Effective gamification is dependent on motivation (Post, 2014), as people can be driven to do something because of internal or external motivation. Gamification is the application of gaming metaphors to real life tasks to influence behaviour, improve motivation and enhance engagement (Marczewski, 2013). Modern videogames use sophisticated psychology and neurochemistry to determine what motivates players to keep returning to the game (Post, 2014). For example, a well-structured recognition system based on a feedback, friends and fun model could create a Las Vegas-style atmosphere in a worker's limbic system, stimulating the amygdala and conditioning the brain to pursue ever-greater acts of sales accomplishment (Strohmeyer, 2013).

Discussing the science and psychology behind gamification, Wu (2011) suggested that the goal of game dynamics is to drive a user-desired behaviour predictably. Hence, it is necessary to understand how humans behave in order to understand game dynamics. According to Choo (2013), a very important factor of gamification is the 'human-focused design' as opposed to the 'function-focused design'. Human-focused design attains that people in the system have feelings, insecurities and reasons why they want or do not want to do things; therefore, it optimises for their feelings, motivations and engagement. Good gamification design does not start with good game elements (mechanics), but it starts with core drives. The gamification framework called Octalysis, introduced by Choo (2013), analysed eight core drives that motivates individuals to do what they do. The Octalysis framework has been discussed in several studies (Ewais & Alluhaidan, 2015; Mora *et al.*, 2015, 2017). These eight core drives are:

(1) Epic meaning and calling: a core drive that keeps the individual motivated because it makes him or her feel that they are something bigger than themselves (e.g. a game gives the individual the opportunity to be the last man standing to save the world when it comes close to an end).
(2) Development and accomplishment: this core drive suggests that individuals are motivated because they feel that they are improving, levelling up or achieving mastery (e.g. Nike+ shows to the individual short-term accomplishments – such as this week's improvement on running one more extra mile than last week – providing the individual the indication that she or he is getting better).
(3) Empowerment of creativity and feedback: this core drive is very similar to LEGO. It gives to users the basic building blocks, so it is up to

him or her to use and utilise their creativity, to try different combinations and strategies, see feedback and adjust, leading to a very engaging process.

(4) Ownership and possession: this core drive suggests that when individuals feel they own something, they want to improve it, protect it and multiply it. It is a core drive that motivates people to accumulate wealth.
(5) Social influence and relatedness: this core drive is based on motives of social influence. It is argued that one of the best ways to change people's behaviour is to show them how others are doing.
(6) Scarcity and impatience: this core drive is based on the notion that people want something just because they cannot have it (e.g. last-minute.com uses the element of rush; when the consumer has the opportunity to see the number of rooms available in a specific category (special offers), then the sense of urgency is increased).
(7) Unpredictability and curiosity: this core drive is based on the element of unknown; as the individual does not know what is coming next, he or she is always thinking about it. This core drive is mostly applied in the gambling industry, but it is the same core drive that leads people to want to finish a book or watch a movie (e.g. Speed Camera Lottery is a reward system where drivers (who keep to the speed limit) have the chance to win money from the other drivers, who have over-exceeded the speed limit. Speed Camera Lottery successfully reduces the speed by 20%, because drivers are willing to slow down (even though their chances of winning are very low), in case that they are actually the winners of the lottery).
(8) Loss and avoidance: this core drive is about doing something to avoid the loss. People do not want bad things to happen. An example of this core drive is Zombies, Run. Unlike Nike+ (which motivates individuals to exercise, to feel accomplish, feel improving and growing), Zombies, Run makes people run and exercise, because they do not want to be eaten by zombies. In this instance, people are motivated to run in order to avoid the loss.

Hence, motivation plays an important part in gameplay, be it intrinsic or extrinsic. Understanding motivation of players is an area in need of further research (Ryan *et al.*, 2006). Considering the fact that motivation is a central topic within gamification and gamified systems are implemented to influence behaviour for wanted and desirable activities, gamification tries to address this intrinsic motivation by applying game design thinking in order to engage people into meaningful and effective activities (Xu *et al.*, 2014). Therefore, it is important when using game elements and game mechanics to identify whether individuals would be motivated to play the game.

Gamers' Profile

To highlight the importance of individuals, Killian (2013), argues that even though it is within human nature to like games, not everyone likes the same kind or style of games. According to Marczewski (2013) even though it is possible to design games, serious games or gamified systems without knowing who the target players and users are, it is more likely to create a more engaging experience when the target players are identified first. He also suggests that it would be more rational to step back and initially view the situation with two players' types in mind; those who are 'willing' to play and those 'not willing' to play (Marczewski, 2013). Key consideration of those willing and unwilling to play is the element of 'fun'. The most important factor that a designer should identify when staring to work on a game is what will be the fun element for the player during the game (Moore, 2011). Since fun is subjective, there are so many different features that must go into a game to make it fun (Dunniway & Novak, 2008).

Richard Bartle's research in the areas of game design and game development, explored players' personality types, for massively multiplayer online (MMO) games. He is best known for his theory on game participant psychology, which classifies players based on their gaming preference (Bartle, 1996). He identified four characteristics of individuals (as players), enhancing the element of fun in specific activities in the game. These four activities are achieving, exploring, socialising and imposing upon others. Most individuals leaned at least a little towards all four, but each player type tended to have a particular overall preference (Bartle, 1996). Moore (2011) further highlights this point claiming that there are actions players perform that are fun, and these are the actions players want to do more often; whereas actions that are not as much fun are only performed as necessary precursors to the fun actions. Bartle's four groups (achievers, explorers, socialisers and killers) seemed to have different preferences and perceptions as to which actions and behaviours are fun and these are described below.

(1) Achievement within the game context (achievers)

In this category, players give themselves game-related goals and vigorously set out to achieve them (Bartle, 1996). For achievers, the main goal is to gather points and rise in levels, and all else is ultimately subservient to this. Exploration is only necessary in order to find new sources of treasure or improving ways of wringing points from it. Socialising is only important as a relaxing method and discovering what other players know about the business of accumulating points. Imposition upon others is only necessary to eliminate rivals or people who get in the way, or even to win vast amounts of points (if there are points awarded for killing other players). Kim (2011) argues that points, badges, levels and leaderboards (the main methodology of gamification), appeals primarily to achiever types, so if an organisation is working with these elements, they will attract this type of gamer.

Achievers would say things like 'I am busy', 'Sure, I will help you. What will I get back?', 'So how do YOU kill the dragon, then?' and 'Only 4211 points to go!' (Bartle, 1996). This group of people would be motivated by the core drive of development and accomplishment, as they are gamers playing the game mainly for the sense of accomplishment and development, and game mechanics (such as points) are more likely to create engagement with them. Furthermore, verbs such as customise, layout and dress up are most appropriate for this category due to the fact that achievers need to develop, customise and dress up their avatar to be ready and prepared to achieve their goals. Hence, it is proposed for a company that focuses on an audience with the characteristics of achievers to include the mechanics of points (since it allows the user to collect points in order to achieve), a progress bar (so the user has a better understanding of progress so far) and tasks (that allow the user to achieve goals).

(2) Exploration of the game (explorers)

In this category, players try to find out as much as they can about the virtual world (Bartle, 1996). Even though initially this means mapping its topology, later it advances to experimentation with its physics. Explorers are delighted in having the game expose its internal machinations to them. Their interest into other mechanics of the game is limited and depends on whether it can help them to find more areas of the game to explore. For them, achieving points is possibly necessary to enter some next phase of exploration, but it is tedious, and everyone can do it. Imposing upon others is quicker and possibly a constructive exercise in its own right, but it causes too much hassle in the long run if the deceased return to seek retribution. Lastly, their interest into socialising develops as an informative source of new ideas to try out, but the real fun comes only from the discovery and making the most complete set of maps in existence. This group of players are individuals who feel successful when they know all the little secrets of the game; therefore, they would play the game hundreds of times if they had to in order to identify these secrets to strengthen the opinion of exploring social engagement (Zichermann & Linder, 2010).

Explorers would say things like 'Hmm…', 'You mean you *don't know* the shortest route from <obscure room 1> to <obscure room 2>?' and 'I have not tried that one, what's it do?' (Bartle, 1996). This group of people would be motivated by the core drive of unpredictability and curiosity as they are gamers looking to explore the unknown initially by mapping its topology, later it advances to experimentation with its physics. Exploration verbs such as read, search, collect and complete are likely to stimulate the interest of explorers. It is proposed to apply tasks that would enhance the element of exploration either in the virtual or physical world to attract and maintain interest for users with the characteristics of explorers. Game mechanics like badges as recognition for their exploration are going to further positively enhance their feelings towards the company.

(3) Socialising with others (socialisers)

This group of players would use the game as a communicative facility and apply the role-playing that it involves as a context in which to converse (and interact) with other players (Bartle, 1996). Socialisers are interested in people and what they have to say. For socialisers, the game is just a backdrop, a common ground where things happen to players. Inter-player activities are very important; empathising with people, joking, listening, sympathising and entertaining, even merely observing people can be rewarding (e.g. seeing them grow as individuals, maturing over time). Exploration could be necessary to understand what everyone else is talking about and achievements could be required in order to get access to features available only to higher levels. Killing is something only ever to be excused if it is absolutely inevitable, or an act of revenge upon someone who has caused difficulty to a friend. The actual fun of the game derives from knowing people, understanding them and forming lasting relationships.

Socialisers would say thing like 'Hi', 'Yeah, well I am having trouble with my boyfriend', 'What happened? I missed it, I was talking' and 'Really? Oh no! Gee, that's terrible! Are you sure? Awful, just awful!' (Bartle, 1996). They are driven by the core epic meaning and calling since this group of people is interested in social engagement of cooperating. Cooperative verbs such as join, share, help, gift, greed, exchange and trade are more likely to attract users' attention. Games do use mechanics to enhance the element of socialising making the game social by giving people the opportunity to share their accomplishments. Thus, it is proposed for companies to encourage this behaviour within their platform either by allowing players to share their accomplishments or even by socialising with each other.

(4) Imposition upon others (killers)

In this group of players use the tools provided by the game to cause distress to other players. Where permitted, this usually involves acquiring some weapon and applying it enthusiastically to the persona (character) of another player in the game world. This group of players enjoy imposing themselves on others. The players in this category disturb other players with a view of killing off their personae, aiming to cause the most stress possible. Achievements and normal point-scoring is only important so as to become powerful enough to begin causing havoc. Exploration is necessary to discover new and ingenious ways to kill people. Socialising is possible to be worthwhile beyond taunting a recent victim (e.g. in finding out someone's playing habits or discussing tactics with fellow killers). For the killers the actual fun of the game comes from the fact that a real person, somewhere, is very upset by what they have caused them.

Killers would say things like 'Ha!', 'Coward' and 'Die!' (Bartle, 1996). Even though this group of users want to achieve in the game, the fact that they do it to promote their superiority against others indicates that the core drive of development and accomplishment is not enough to explain their behaviour. Instead it is likely to assume that they fall under the core drive of social influence and relatedness since they are users attracted by comparing themselves with others and thus promoting their superiority. Competitive verbs including win, beat, brag, taunt, challenge and pass are all verbs of competitive activities that encourage their behaviour. Even though in games Bartle (1996) mentions that they are important for the balance of the games in a gamified system, their toxic behaviour can also be negative for the balance of the game. Game mechanics such as a leaderboard is more likely to attract users with these characteristics, since it allows them to be compared with others and promote their superiority, but at the same time it removes their toxicity.

However, not all research about digital games can be applied directly to the gamification of other applications (Dixon, 2011). Games are just for fun and entertaining, whereas gamification has a certain purpose, without any gameplay (Lombriser & van der Valk, 2011). These differences possibly lead to the different players' profiles between gamers and gamification users.

Marczewski (2013) suggested that Bartle's player types may not translate well to non-MMO games or any other non-game-related systems such as gamified systems. He wanted to look at a more gamification-specific taxonomy for types of users within systems (Parapanos & Michopoulou, 2019). Therefore, starting from the perspective of intrinsic motivation, he chose to use ideas from Edward Deci and Richard Ryan's self-determination theory and Daniel Pink's three motivational drives (Pink, 2010). Marczewski (2013), by combining these two theories, suggested four motivations: (1) relatedness (to desire, to socialise); (2) autonomy (the urge that an individual has to direct his or her own life); (c) mastery (the desire to get better and better in something that matters); and (4) purpose (the force to do an activity in the service of something with bigger meaning). These four intrinsic motivations were used to describe four intrinsically motivated user types: socialisers, free spirits, achievers and philanthropists, respectively. However, a fifth type was suggested; an extrinsically motivated type, the player, who is actually motivated by the reward.

Players are happy to 'play' the game, where points and rewards are up for collection. Finally, and much later on, a sixth type was included, the disruptor. The six types gamified system users are described below.

(1) Socialisers

Socialisers are motivated by relatedness and they want to interact with others and create social connections (identical to one of Bartle's types of players, the socialiser) (Marczewski, 2014). They are interested in parts of the system that enable them to accomplish this, and they promote and evangelise the internal social network. This group values the social engagement of cooperating. According to Kim (2011), cooperative verbs include join, share, help, gift, greed, exchange and trade. Suggested design elements for this group are guilds or teams, social networks, social comparison, social competition and social discovery. An example of social competition could be promoted by leaderboards. Leaderboards work in order to motivate and encourage users to become players. For example, Foursquare started with city-centric leaderboards, but now places the emphasis on ranking users against their friends. Earn a few points for a check-in, and Foursquare will show the user which of his or her friends they would overtake on the leaderboard. In a hotels gamified application, one massive leaderboard could be a negative factor for users, since it would be very difficult to reach the top five or 10 with the biggest rewards. In this case, social comparison will not benefit the system. However, a complex system with local leaderboards at each destination with unique rewards during their visit at a hotel could be better. This mechanic, applied with the right tasks, will increase the social interaction between users or even the locals. Furthermore, if on reaching the top 10 at those local easily accessed leaderboards, reward items that contribute towards a bigger global leaderboard with the biggest rewards, could be an increased motivator. Social media as internet-based applications allow the creation and exchange of user-generated content, reshaping the way tourism-related information is distributed and the way people make plans to travel (García-Pablos *et al.*, 2016); thus, they are used by organisations to gamify their strategy. For example, Caesars Casino provides the individual the opportunity to earn reward credits through gaming, but also through social gaming by playing Caesars Interactive Facebook games. Total rewards members use the social rewards programme to engage with the brand on Facebook, Twitter and YouTube to earn even more credits. These twists on the typical points-based approach are making the total rewards an even more effective loyalty driver. Even though this example is very similar to the 'Jet Off Geneva' and the game mechanic of leaderboard, it also includes other game mechanics such as points and levels, appointments, progress and sharing. Furthermore, Caesars interactive Facebook and social rewards programmes give the ability to the individual to socialise. This gamified system offers the individual empowerment of creativity and feedback, social influence and relatedness.

(2) Free spirits

Free spirits are motivated by autonomy and they want to create and explore (Marczewski, 2014). Choo (2013) adds that these individuals are not willing to be restricted in how they go through their personal journey. They are the most creative users having the fanciest avatar, they create the most personal content, but they often find the defects of the system. Free spirits have identical characteristics with explorers (as identified by Bartle), since they have the intrinsic need of self-expression by exploring the game in order to get the most from the system for their own enjoyment. This description agrees with the unpredictability and curiosity core drive, given earlier to explorers. However, they also show characteristics based on the core mechanic of empowerment, creativity and feedback. Even though the user is interested in exploring the system in a similar way to a gamer, users in this case are also attracted by the freedom that the platform is promoting to empower their creativity to try different combinations. Suggested design elements for this group are exploratory tasks, Easter eggs, unlockable content, creativity tools and customisation. In a hotel's gamified system, a badge system would increase the users' motivation in completing tasks. Each badge should represent an achievement mastered by travelling guests, with emphasis on the level of commitment. Badges such as veteran, star or governor when users prove that they have developed mastery of certain tasks are good ideas as they establish their superiority in the system. Gamification during tourists' holidays is associated with learning about the environment. Tourists often want to visit specific points of interests (PoIs) such as historical buildings, unique landscape or even haunted places (Linaza *et al.*, 2014). Gamification in this context can apply game mechanics in order to create fun and learning elements. In such games, tourists not only follow a list of recommendations given by a mobile application, but also learn something about their environment by solving mini-games related to their experience (Linaza *et al.*, 2014). Such an example is presented by Linaza *et al.* (2014) within the ExCORA, which is the battle between the French and Anglo-Portuguese troops that occurred in San Sebastian from 28th June to 8th December 1813. The game takes place in the Urgull Mountain where the fortress was placed and the players walking around would have to find several relevant locations so that they can interact with different elements. The game starts outside the walls of the fortress and finishes with the flag located on the top of the Urgull Mountain. Players take the role of an English soldier tying to defeat the French army garrisons in the forest and the experience has six PoIs and four treasures with different contents, validation methods and related mini-games (Linaza *et al.*, 2014).

One online mobile gaming application developed for tourists during their visit to the destination of Stockholm is 'Stockholm Sounds'. The application is being launched in order to offer tourists unique experiences that usually could not be found in guidebooks. According to Swedania (2013), tourists are challenged to discover Stockholm through game missions, interactive experiences and visits to exciting places based on the sounds and music of the city. In particular, as soon as passengers land at Stockholm Arlanda and open the application, they receive suggestions for experiences and a map to help them find the places at the airport. For example, in the baggage hall, people can answer questions about some of the musicians found in the Stockholm hall of fame. Each correct answer and each experience gives them one point. Visitors can also share their experiences from different game activities with friends, which creates opportunities for viral spread (Swedania, 2013). This gamified system promotes the element of fun, through a form of gameplay, but also the tourist (as gamer) has the opportunity to learn more about Stockholm and music, while at the same time has the opportunity to win prizes for completing tasks. Indeed, by visiting over 40 locations in Stockholm, tourists earn 'points' by completing challenges, history quizzes or gaming sessions, all connected to music or sound experiences; for example, from discovering 'the sound of sushi' at Akki Sushi in Södermalm, in a music quiz at popular record store Pet Sounds. Completing the mini-tasks unlocks events and rewards including pre-party playlists on Spotify and information about hangouts around the city that are popular with the locals (Liew, 2013). This gamified system includes core drives such as epic meaning and calling, development and accomplishment (by fulfilling the tasks), social influence and relatedness (by sharing activities with friends), and unpredictability and curiosity (by giving the opportunity to the tourist to discover Stockholm within the game application).

(3) Achievers

Achievers are motivated by mastery and they are looking to learn new things and improve themselves by overcoming challenges. These individuals want to be perfect on the internal learning system. Even though they do this for their own satisfaction, they do not mind showing off. The difference from achiever as has been presented by Bartle (1996), is that achievers' main goal is to gather points and rise in levels; all is ultimately subservient to this. However, they are not that interested in showing off and only interact with killers in cases where they will be rewarded with more points. Achievers in Marczewski's categorisation may also be motivated by status as a result of their personal achievement. Hence, users' behaviour is explained by a combination of development and accomplishment core drive and social influence and relatedness core drive. The complexity of the behaviour might be a result of the core difference of the system itself. In a gamified system achieving itself might not be enough of a core drive to motivate the user to return to the system, since the reward will possibly be associated in a competition with other users. Suggested design elements for this group are challenges, certificates, learning new skills, quests, levels or progression and epic challenges. An example of a gamified strategy based only on game mechanics, such as points and leaderboards, is Jet Off Geneva. Jet Off Geneva is an online game that gives the individual (player) the opportunity to play and win by pumping the most litres of water and finishing first on the leaderboard. The system promotes an element of fun as it includes a gameplay experience, but it is lacking on crucial elements of gamification and game thinking such as epic meaning and calling, development and accomplishment, empowerment of creativity and feedback, ownership and possession, social influence and relatedness, scarcity and impatience, unpredictability and curiosity, loss and avoidance. Instead, a quest attached to the history of the destination could motivate the visitor to get involved and explore the area. Each quest could be turned in only once and before the visitor leaves the destination. These could increase the commitment of the user (with the profile of an explorer) with the system at the destination, making this interaction more interesting and fun. Still, the system should use hints in case the explorer gets stuck on the quests in order to avoid frustration.

(4) Philanthropists

Philanthropists are motivated by purpose and meaning. This group of individuals are altruistic, wanting to give to other people and enrich the lives of others in some way, with no expectation of reward (Marczewski, 2014). Philanthropists feel that they are part of something greater and want to give back to others. This group of users clearly enjoy being part of something greater and want to give back to others deriving from the epic meaning and calling core drive. A gamified application can help them achieve their inner goals by gifting people who need a tangible or intangible good or service. They offer selfless dedication for 'the cause', because they enjoy helping. Suggested design elements for this group are collection and trading, gifting, knowledge sharing and administrative roles. This group of users could use the application for altruistic purposes. Altruism represents an individual's willingness to benefit the wellbeing of others on a voluntary basis without the anticipation of any form of return (Chen *et al.*, 2013). The ability of the application to be used as a gift tool would benefit users with these characteristics; for example, booking a weekend for friends or family would meet altruistic purposes.

(5) Players

Players are motivated by rewards and they do what is needed of them in order to collect rewards from the system (Marczewski, 2014). They are in it for themselves. The difference between players and any other category (achievers, philanthropists, free spirits and socialisers) is the fact that they are extrinsically motivated by the rewards and not by any other intrinsic motivation such as relatedness, autonomy, mastery or purpose. This group of individuals like to receive the achievements in the system and have their names in the leaderboards. The fact that they want to achieve in the system is not the result of the intrinsic motivation of mastery, but the reward itself. They are happy to take advantage of 'loopholes' to gain an edge. This is the only group of individuals in gamified systems that are willing to play the game for extrinsic factors (rewards). This is the most likely group of players to be attracted and engaged in a hospitality and tourism gamified system (Choo, 2013). It is argued that reward-based gamification is suitable for some situations (Nicholson, 2015). For example, when the organisation is looking for immediate and short-term change, reward-based gamification can certainly create that (Nicholson, 2015). Many reward-based gamification systems create an immediate spike in engagement as users strive to explore this new system. As long as the organisation is willing to continue supplying rewards, the behaviours can continue by those motivated to earn the rewards (Nicholson, 2015). To apply bonuses in gamified Information Systems, they must fit to the underlying reward system (Thiebes *et al.*, 2014). Thereafter, the purpose of Information System is to turn raw data into useful information for future decision making within an organisation. For instance, bonus points can be rewarded after successful completion of a special task or achievement (Thiebes *et al.*, 2014) Another example might be a tourism organisation that could reward the user with points or badges to acknowledge the achievement of visiting a specific amount of destinations, encouraging the individual to be attached with the system. Moreover, bonus mini games may be awarded in Information System, after a completion of a series of tiring tasks, aiming to re-establish concentration and motivation (Thiebes *et al.*, 2014). In this case, tasks associated with the history of the destination applied with the appropriate reward (either tangible, such as discounts, or intangible, such as points and badges) is likely to encourage the user to visit the local area and stay engage with the system. A gamified system by The InterContinental Hotels Group (IHG) uses a trivia game (Win it in a Minute) that awards correct answers with free miles and capitalises on the universal desire to prove how smart we are. It is also proved a smart move because the vice president of loyalty programmes reported that in the first two weeks of the Win It in a Minute promotion IHG has seen 100,000 game plays and has handed out more than 100 million priority club points. This gamified system is based on the rewards since users' correct answers are immediately transformed into free miles.

(6) Disruptors

Disruptors are motivated by change. They want to disrupt the system, either directly or through other users to force positive or negative change (Marczewski, 2014). This group of people seems to have some identical behaviour with killers from Bartle's (1996) typology. This category of players likes to compete. Even though killers have been linked with the core drive of social influence and relatedness, disruptors appear to have more characteristics that are similar to the ownership and possession core drive. This is explained due to the fact that killers are interested in showing off their superiority without damaging the system itself, whereas disruptors express behaviour negative to the system itself. They feel they own the system and disturb it based on their personal opinion. Large numbers of users from this category would mostly have a negative impact on the system, in interactions with others, rather than positive. Even though a small number of players fit these characteristics, they are quite important for socialisers (the biggest group), because without the killers, socialisers would have little to talk about. In the same line of thought, for a gamified system, disruptors can indirectly benefit the system as they could give socialisers something to talk about or philanthropist something to give back to others.

A gamified system should be balanced for all users. Choo (2013), describing a balanced gamified system, suggests that it should exclude the sixth user type (disruptor) and simplify the approach by using the other five user types (player, philanthropist, achiever, socialiser and free spirit). Furthermore, the system has to be appealing to the four basic intrinsic motivations as categorised by Bartle (1996) (exploring, achieving, socialising and imposing upon others) and user types (make it social and meaningful) as well as giving freedom to the users. Once this structure is established, a thoughtful system of rewards can be integrated. It is important for the system not to be dependent on the rewards to function effectively. It is also imperative for the system not to have too many users with extrinsic behaviour (for whom rewards are the only important element), as the system will run the risk of devaluation. In contrast, philanthropists and achievers could both help a system thrive. Philanthropists try to help everyone, contributing to in any way possible. Achievers may act the same depending on the situation, demonstrating their abilities rather than actually helping others (Marczewski, 2014). Socialisers, even though they do not offer as much as other types with respect to internal contribution, are effective for evangelising and bringing others to the system (Marczewski, 2014). Therefore, the point of identifying the gamification user types is to provide a better understanding of why and how people would use a system, but also how it could create better engagement with them (Marczewski, 2014).

There are significant differences in profiling individuals from playing a game to being gamified application users. In a gamified situation, individuals will not necessarily be able to have the same freedom to 'play' and 'explore' the game. This occurs because games and gamification are designed from different perspectives. There has been some confusion about games and gamification, but they are two very different entities (Post, 2014). The primary purpose of games is to entertain, whereas gamification seeks to motivate people to change behaviours, develop new skills or engage in innovation. Therefore individuals want different things out of gamification and games (Post, 2014). Hence, as gamification applications differ from ordinary games, thus the game mechanics used should be carefully chosen and adapted for the purpose.

Gamers' Life Cycle

Player types have been recognised and divided according to their preferences in either playing a game (Bartle's types of players) or being gamification users (Marczewski's users' types), based on the needs of individuals from a game or gamified application. Kim (2011) however, looked into the player life cycle. For Kim (2011), players have to be approached with different techniques in each stage of the game, according to the frequency that they visit the game. There are three stages that are actually in order and normally these stages cannot be jumped, according to Kim (2011). The first stage is called boarding for newbies and it is the stage where the game successfully introduces the gameplay to gamers. The second stage, habit-building, is the phase that builds all the elements that could make the game a daily or weekly habit. Finally, stage three is called mastery and is based on the exclusivity that the players get from the game. The issue is that gamers have different needs in different stages. Similarly organisations have to think about these stages when designing a gamified application so as to build a sustainable social system.

Kim (2011) highlights the importance of learning for the life cycle stages of gamers, as each one of them has different focus of learning. For example, newbies need to learn the basics, so to encourage individuals not only to think about their first visit, but the first couple of months of experience. They get the feeling that they know how the game works and how to move on. Then regulars are people who have learned the basics and they are engaged, but they have to keep being re-engaged. The most important thing is to introduce new content activities or challenges, so they need something to keep them coming back. For example it could be other people (in the case of socialisers), or it could be new challenges (in the case of achievers). Even though points and levels help individuals to come back for more, it is a series of innovations that keeps individuals coming back. Lastly, enthusiasts or experts need something that it is exclusive to them, such as a form of recognition and impact that they could have in the world (Kim, 2011).

The gamers' life cycle may include different stages and characteristics as illustrated by two examples elaborated below. These two examples explain how the context and type of game yield different types of user behaviour. The examples are drawn from some of the most popular types of games; namely, MMORPGs (massively multi-player online role-playing games) and RTS (real-time strategy).

Jiang (2008) identified four stages in the players' life cycle after carrying out research in two MMORPGs (Mir II and World of Warcraft). MMORPGs are a little different from pure action games as they usually add a dimension to what otherwise is a game involving little more than wandering around shooting things (Lecky-Thompson, 2009). There is a sense of travel and progression that is not necessarily prevalent in adventure or combat games, and is generally a result of the role-play aspect of the game, in part because actions taken ex game are almost as important as those as taken in game (e.g. customising the player and interactions that occur in the player forums) (Lecky-Thompson, 2009). Key points from the players' point of view include ownership (of objects or locations), levelling up (to improve capabilities), revenue model (advertising and upgrades) and in-game currency economy; therefore, unlike in a pure arcade-style gaming model, players are generally encouraged to empathise more with their persona. With regard to the revenue model, they tend to be long term, ranging from requiring payment to keep a player's account in good standing to pushing players to buy expansion packs and client upgrades (Lecky-Thompson, 2009).

Thus, the gamers' life cycle can be divided in four stages: confusion, excitement, involvement and boredom (Jiang, 2008). Confusion (0–1 months) is the stage where new players receive detailed information. If they receive proper guidance during this period, gamers will stay and enter the next period, otherwise they will soon leave (Jiang, 2008). Excitement (2–4 months) is the period full of enhanced chat, guilds, rewards for exploring, skills growth and clear growth plans. Involvement (2 months to 4+ years), is the longest period of a gamer's life cycle, filled with creating social events, massive ability, quests, events and new land. Lastly, boredom (2–4 months) is the period when, if there is nothing new or exciting after involvement, the gamers leave. In different stages of gamers' life cycle, they focus on different characteristics of the game. In the early stages, they focus more on features and entertainment, but later on their interest is more on socialisation and relationships. Hence, gamers' profiles change depending on the time spent on the game. A gamer might start a game as an achiever and finish it as a socialiser.

Fischer (2014) identified five phases that develop players' life cycle and these are the initiation phase, competence phase, conceptual mastery, training phase and dominance phase. Fischer (2014) identified these five phases of players' life cycle after research into games referred to as RTS games or turn-based games. RTS games are different from other real-time in action games in that they are almost turn-by-turn in execution

(Lecky-Thompson, 2009). Some of the most popular RTS games tend to strictly be turn based, which means each round consists of one move per player, with the game unable to progress until all players have made their move (Lecky-Thompson, 2009). Within this game category, players tend to play locally, with other players represented as characters within the local game and only the actions (or even just the results of the actions) of each player are communicated, via a server to the clients.

The initiation phase is the stage that the player learns and internalises the rule set. Fun, engagement and intellectual value are steadily, however rather slowly, increasing as more and more potential possibilities of interaction are discovered. The competence phase is the stage where the players learn what 'good' and 'bad' actions mean in the context of the game's goal, and how to identify them. The player is constantly getting better at mastering challenges and reliably manipulating the system to his or her own advantage. Fun and intellectual value reach their maximum. Conceptual mastery is the stage where the player has essentially mastered the system. The ultimate 'skill ceiling' (phase 5) is in sight and the necessary steps to get there (phase 4) are already quite clearly perceptible. The training phase is the stage where the player is, through repeated exercise, approaching the limits of skill in the game. This can, for example, consist of memorising optimal actions in specific, already completely solved situations, or perfecting the reliability of the performance in action games. The learning process is less rapid than in phase 2, getting better takes more effort and the potential value is dwindling. Finally, the dominance phase is the stage where the player has reached the skill ceiling and thus fully exhausted the game's depth. There is nothing more to discover or to learn. The game is solved and has therefore lost its intellectual value, and with that the ability to provide fun.

The examples of Jiang (2008) and Fischer (2014) highlight that players' life cycle is different according to game design characteristics. In this line of thought and following Kim's (2011) argument that players have to be approached with different techniques in each stage of the game in order to build a sustainable social system for the organisation, the game mechanics have to be carefully chosen according to the player's profile and the phase they are in the game.

Conclusion

Tourism destinations, as well as hospitality companies, use interactive games in their web pages, blogs or social networks (Firoiu & Croitoru, 2014). Tourism and hospitality industries have long been recognised as some of the most 'globalised' industries in the service sector (Qi, 2016). Even though the internet has been a key factor in this globalisation process, and hotel websites have become indispensable (Qi, 2016), gamifying a website is not a simple thing to do. Gamers do not play a game for collecting points or badges, but because games are challenging, allow them to use their creativity, socialise or even be something more than what they are in real life. Therefore, game mechanics are not enough in themselves to attract an individual to play a game; they are only tools to support building it. It is important for the success of the system to focus on the needs of the audience before applying the appropriate game mechanics.

Gamification is used by brands to motivate employees, create healthy competition among teams, generate buzz or social proof and encourage customer loyalty (Stanley, 2014). This has also been evident within the tourism and hospitality industry, where gamification has been seen to encourage tourist engagement, enhance tourist experience, improve tourist loyalty and increase tourism brand awareness (Killian, 2013; Xu *et al.*, 2014) applied in technology-mediated and non-technology-mediated contexts (Bulencea & Egger, 2015). A tourist is a person who typically has 'little or no knowledge of the environment' (McKercher & du Cros, 2003). Using gamified systems allows the tourist to learn more about the local environment by having fun. Gamification enables destinations and hotels to promote the history of the location and create an emotional experience between the tourist and PoI (Xu *et al.*, 2015). Gamification should enable tourists and hotel visitors as gamers to have a sense of engagement, immediate feedback, a feeling of accomplishment and the success of striving against a challenge and overcoming it (Kirsh, 2014).

Gamification is implemented with a variety of techniques (some easy to implement, some requiring advanced planning, coding or technical expertise), so any business may use gamification to achieve better results, no matter what the goals are (Stanley, 2014). Many gamified applications are failing due to poor game design. Some organisations focus on the obvious game mechanics, such as points, badges and leaderboards, rather than the subtler and more important game design elements, such as balancing competition and collaboration, or defining a meaningful game economy (Burke, 2013). However, interest should be taken into understanding the motives of users when using the system in order to make it more attractive and engaging. Bartle's taxonomy of players is a great example from the gaming industry in understanding the needs and wants of gamers when playing the game. Marczewski's taxonomy of users highlights the importance of identifying the characteristics of the users in order to enhance the engagement in using the system. Gamification is not fundamentally wrong; it just needs to evolve past the hype. In conclusion, gamification is a complex phenomenon that we are only just beginning to understand. Further research is required to further conceptualise and comprehend the intricacies inherent in the definitions and meanings currently ascribed to it. In particular, the use of game mechanics and game thinking as they relate to user profiling needs further investigation to uncover the core constructs that affect the intention to use gamified applications.

References

Al-Zaidi, Z. (2012) Gamification's march to ubiquity. *The Guardian*. Available at: www.theguardian.com/media-network/media-network-blog/2012/apr/26/gamification-ubiquity (accessed August 2018).

Bartle, R. (1996) Hearts, clubs, diamonds, spades: Players who suit MUDs. Available at: http://mud.co.uk/richard/hcds.htm (accessed February 2018).

Buckner, C. (2014) How to effectively use badges in your business. Available at: www.gamification.co/2014/10/30/how-to-effectively-use-badges-in-your-business/ (accessed November 2018).

Burke, B. (2013) How gamification motivates the masses. *Forbes*. Available at: www.forbes.com/sites/gartnergroup/2014/04/10/how-gamification-motivates-the-masses/ (accessed January 2018).

Chen, H.L., Fan, H.L. and Tsai, C.C. (2013) The role of community trust and altruism in knowledge sharing: An investigation of a virtual community of teacher professionals. *Educational Technology and Society* 17 (3), 168–179.

Choo, Y-K. (2013) What is gamification. Available at: http://www.yukaichou.com/gamification-examples/what-is-gamification/#.Uwyrnvl_u6M (accessed February 2018).

Deterding, S., Dixon, D., Khaled, R. and Nacke, L. (2011) From game design elements to gamefulness: Defining gamification. Paper presented at the 15th International Academic MindTrek Conference: Envisioning Future Media Environments, 28–30 September, Tampere, Finland.

Dickey, M. (2005) Engaging by design: How engagement strategies in popular computer and video games can inform instructional design. *Educational Technology Research and Development* 52 (2), 67–83.

Dixon D. (2011) Player types and gamification. Available at: http://gamification-research.org/wp-content/uploads/2011/04/11-Dixon.pdf (accessed February 2018).

Dominguez, A., Saenz-de-Navarrete, J., de-Marcos, L., Fernandez-Sanz, L., Pages, C. and Martinez-Herraiz, J-J. (2013) Gamifying learning experiences: Practical implications and outcomes. *Computers and Education* 63, 380–392.

Dunniway, T. and Novak, J. (2008) *Game Development Essentials: Gameplay Mechanics*. New York: Delmar Cengage Learning.

Bulencea, P. and Egger, R. (2015) *Gamification in Tourism: Designing Memorable Experiences*. Norderstedt: Books on Demand.

Ewais, S. and Alluhaidan, A. (2015) Classification of stress management mHealth apps based on Octalysis framework. Paper presented at the 21st Americas Conference on Information Systems, 13–15 August, Puerto Rico.

Firoiu, D. and Croitoru, A.G. (2014) Tourism and tourism infrastructure from the perspective of technological changes. *Romanian Economic and Business Review* 8 (2), 93–103.

Fischer, F. (2014) Why elegance matters: The lifecycle of games. Available at: www.gamasutra.com/blogs/FabianFischer/20141008/227318/Why_Elegance_Matters_The_Lifecycle_of_Games.php (accessed November 2018).

García-Pablos, A., Lo Duca, A., Cuadros, M., Linaza, M.T. and Marchetti, A. (2016) Correlating languages and sentiment analysis on the basis of text-based reviews. In A. Inversini and R. Schegg (eds) *Information and Communication Technologies in Tourism* (pp. 565–577). Switzerland: Springer.

Hanus, M.D. and Fox, J. (2015) Assessing the effects of gamification in the classroom: A longitudinal study on intrinsic motivation, social comparison, satisfaction, effort, and academic performance. *Computers and Education* 80, 152–161.

Huotari, K. and Hamari, J. (2012) Defining gamification – A service marketing perspective. *MindTrek* 3 (5), 17–22.

Jiang, Q. (2008) Gamers' lifecycle and narrative structure: The comparison of online gamers' experiences in the Legend of Mir II and World of Warcraft. *China Media Research* 4 (2), 26–31.

Killian, E. (2013) *Gamification 2.0 – A Concept*. Available at: http://books.google.co.uk/books?id=wJJbAgAAQBAJ&pg=PA16&dq=gamification+definition&hl=en&sa=X&ei=Wk9iU-XVNIXGPP-NgLgK&ved=0CF0Q6AEwAw#v=onepage&q=gamification%20definition&f=false (accessed 17 December 2020).

Kim, A.J. (2011) Smart gamification: Seven core concepts for creating compelling experiences. Available at: https://www.youtube.com/watch?v=F4YP-hGZTuA (accessed 12 November 2018).

Kirsh, B.A. (2014) *Game in Libraries: Essays on Using Play to Connect and Instruct*. North Carolina: McFarland & Company Inc. Publishers.

Lecky-Thompson, G.W. (2009) *Fundamentals of Network Game Development*. Boston: Course Technology.

Liew, J. (2013) New app offers tourists a 'sonic' Stockholm guide. Available at: www.thelocal.se/20130913/50224 (accessed September 2018).

Linaza, M.T., Gutierrez, A. and Garcia A. (2014) Pervasive augmented reality games to experience tourism destinations. In X. Zheng and I. Tussyadiah (eds) *Information and Communication Technologies in Tourism* (pp. 497–510). Vienna: Springer.

Lombriser, P. and van der Valk, R. (2011) *Improving the Quality of the Software Development Lifecycle with Gamification*. Berlin: Springer-Verlag.

Maedche, A., Botzenhardt, A. and Neer, L. (2012) *Software for People Fundamentals: Trends and Best Practices*. Heidelberg: Springer.

Marczewski, A. (2013) *Gamification: A Simple Introduction*. (2nd edn) (self-published on Amazon Digital Services). Kindle edition, Loc 662 of 1798.

Marczewski, A. (2014) *Marczewski's Gamification User Types*. Available at: http://elearningindustry.com/marczewski-gamification-user-types (accessed February 2018).

McKercher, B. and du Cros, H. (2003) Testing a cultural tourism typology. *International Journal of Tourism Research* 5 (1), 45–58.

Moore, M.E. (2011) *Basics of Game Design*. Boca Raton, Florida: Taylor and Francis Group.

Mora, A., Riera, D., Gonzalez, C. and Arnedo-Moreno, J. (2015) A literature review of gamification design frameworks. Games and Virtual Worlds for Serious Applications (VSGames), 7th International Conference on 2015 (pp. 1–8). Skovde, Sweden: IEEE.

Mora, A., Riera, D., Gonzalez, C. and Arnedo-Moreno, J. (2017) Gamification: A systematic review of design frameworks. *Journal of Computing in Higher Education* (29), 516–548.

Nicholson, S. (2015) A RECIPE for meaningful gamification. In T. Reiners and L.C. Wood (eds) *Gamification in Education and Business* (pp. 1–20) New York: Springer.

O'Brien, J. (2014) Business travel gamification rewards both employees and companies. *Forbes*. Available at: www.forbes.com/sites/united/2014/02/12/business-travel-gamification-rewards-both-employees-and-companies (accessed February 2018).

Parapanos, D. and Michopoulou, E. (2019) Understanding key motivations for using a hotel gamified application. In J. Pesonen and Neidhardt (eds) *Information and Communication Technologies in Tourism* (pp. 411–422). Vienna-New York: Springer.

Pink, D.H. (2010) *Drive: The Surprising Truth About What Motivates Us*. New York: Riverhead Books.

Post, R. (2014) Game on: Could gamification help business change behaviour? *The Guardian*. Available at: www.theguardian.com/sustainable-business/game-on-gamification-business-change-behaviour (accessed January 2018).

Qi, S. (2016) Hotel website quality versus star level: The case of Macau Hotel websites. In A. Inversini and R. Schegg (eds) *Information and Communication Technologies in Tourism* (pp. 325–338). Switzerland: Springer International Publishing.

Ryan, R.M. and Deci, E.L. (2000) Self-determination theory and the facilitation of intrinsic motivation, social development, and well-being. *American Psychological Association, Inc* 55 (1), 68–78.

Ryan, R.M., Rigby, C.S. and Przybylski, A. (2006) *The Motivational Pull of Video Games: A Self-Determination Theory Approach*. New York: Plenum.
Stapleton, A. (2004) Serious games: Serious opportunities. Australian Game Developers Conference, Melbourne, Victoria, Australia, Academic Summit 2004.
Strohmeyer, R. (2013) Gamification: Using play to motivate employees and engage customers. Available at: www.pcworld.com/article/2047564/gamification-using-play-to-motivate-employees-and-engage-customers.html (accessed October 2018).
Swan, C. (2012) Gamification: A new way to shape behaviour. *Communication World* 29 (3), 13–14.
Swedania (2013) Discover Stockholm with the Stockholm Sounds app. Available at: www.swedavia.com/about-swedavia/news/discover-stockholm-with-the-stockholm-sounds-app (accessed August 2018).
Thiebes, S., Lins, S. and Basten, D. (2014) Gamifying information systems: A synthesis of gamification mechanics and dynamics. Paper presented at the European Conference on Information Systems (ECIS), 9–11 June, Tel Aviv.
Van Grove, J. (2011) Gamification: How competition is reinventing business, marketing & everyday life. Available at: http://mashable.com/2011/07/28/gamification (accessed November 2018).
Warmelink, H. (2014) *Online Gaming and Playful Organization*. New York: Routledge.
Wu, M. (2011) Gamification 101: The psychology of motivation. Available at: https://community.lithium.com/t5/Science-of-Social-Blog/Gamification-101-The-Psychology-of-Motivation/ba-p/21864 (accessed August 2018).
Xu, F., Weber, J. and Buhalis, D. (2014) Gamification in tourism. In Z. Xiang and I. Tussyadiah (eds) *Information and Communication Technologies in Tourism* (pp. 525–537). Wien: Springer.
Xu, F., Tian, F., Buhalis, D. and Weber, J. (2013) Marketing tourism via electronic games: Understanding the motivation of tourist players. In: 5th International Conference on Games and Virtual Worlds for Serious Applications (VS-GAMES), 2013, Poole, United Kingdom (pp. 1–8). IEEE.
Zichermann, G. and Cunningham, C. (2011) *Gamification by Design: Implementing Game Mechanics in Web and Mobile Apps*. Newton, MA: O'Reilly Media.
Zichermann, G. and Linder, J. (2010) *Game-Based Marketing: Inspire Customer Loyalty Through Rewards, Challenges, and Contests*. Hoboken, NJ: Wiley.

3 Understanding Games and Gamified Experiences: The MAPS-AIM Model

Russell B. Williams

Introduction

Games come in all shapes and sizes. There are board games, card games, game shows, Olympic games, sports, videogames, computer games, mobile games and real-world games with and without technological augmentation of reality. Games can inspire, inform, persuade and share as well as entertain and provide a platform for social interaction. A game is a struggle towards a goal, whether it happens in a physical space or in cyberspace. In addition, games are imbedded with story, media, means of presentation, underlying structure, categorisation, subjective-experience, audience, intention and measure of success (Adams, 2010; Salen & Zimmerman, 2004; Schell, 2008).

The aim of this chapter is to reconcile and clarify two well-known conceptualisations of games, MDA (mechanics, dynamics and aesthetics) (Hunicke *et al.*, 2004) and MAST (mechanics, aesthetics, story and technology) (Schell, 2008) and expand the *resulting* list of components and enable the application of those terms to gamified experiences in tourism. Making a good game starts with intentionality, the knowledge of what makes a game a game, the ways in which people enjoy games and the experience designed into the game (Costikyan, 2002). These are foundational in the MAPS-AIM model. Games and gamified experiences are made of more than rules, goals and a pleasurable experience. Those components need to be identified and defined.

Mechanics are the one element that both MDA and MAST share (aesthetics is used in MDA to talk about subjective experience rather than a consideration of design and beauty applied in MAST). It is certainly the most obvious element that separates a game from other forms of media and leisure. Mechanics are also a key point of discussion within a number of chapters in this book. Mechanics are essentially the scoring mechanisms

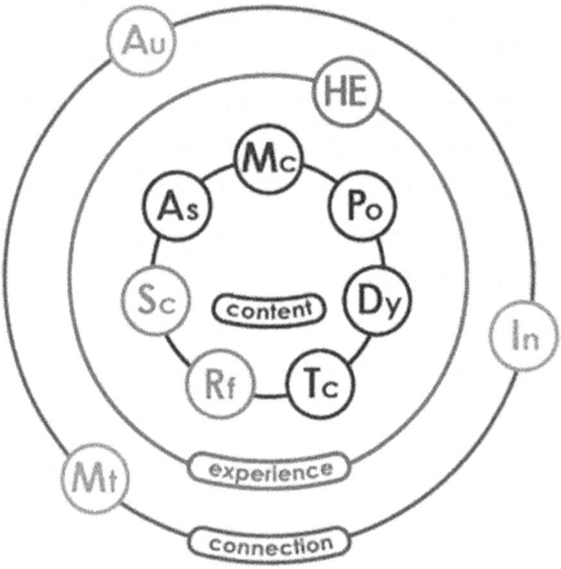

Figure 3.1 MAPS-AIM model. Mc, mechanic; As, aesthetic; Po, poetic; Sc, schematic; Tc, technologic; Rf, referent; Dy, dynamic; HE, hedonic-eudaimonic (HE); Au, audience; In, intention; Mt, metric

that are applied to an environment, activity or process and that underlie gamified experiences. While they are a unique factor, mechanics are not the only element distinguishing games and gamified experience. There are 10 other elements that should also be considered when creating a game or turning touristic and leisure activities into a game.

In the MAPS-AIM model, there are 11 elements of a game: aesthetic, poetic, schematic, technologic, referent, and dynamic which are content elements, experience element, hedonic-eudaimonic experience element, and audience, intention and metric which are connection elements. These will be discussed further (see Figure 3.1).

Games and Play

The starting point for understanding the components of gamified experiences is to define what is a game and what is play. Zimmerman (2004) defines games as voluntary, interactive, controlled by rules, possessing a quantifiable outcome and an artificial conflict. He states that play comes out of submission to the rules, while arguing that rules and play are opposites. ‘Rules are fixed, rigid, closed, and unambiguous. Play on the other hand is uncertain, creative, improvisational, and open-ended. The strange coupling of rules and play is one of the fascinating paradoxes of games' (Zimmerman, 2004: 155). Bateman (2009b) and Despain (2013) give emphasis to Huizinga's (1955) magic circle, the limited space of the game, saying that games are separate from daily life. Adams (2010) calls rules, goals, play and pretending the essential elements of a game. Costikyan (2002: 24) calls a game 'an interactive structure of endogenous meaning that requires players to struggle toward a goal'. Interactive structure in this definition points to the connection between the player, decision-making and the rules of the game. Endogenous meaning is the fact that things in the game only have value and meaning in the game world or within the magic circle of the game. Struggle implies the conflict between players or between players and elements of the game. The goal is what gives all of the struggle and decision-making value.

Looking deeper into the elements of a game Hunicke *et al.* (2004) established the idea that mechanics, dynamics and aesthetics make up the fundamental elements of a game (MDA). Within their scheme mechanics are the rules of play, dynamics are the category of play and aesthetics are the subjective experience of play. Schell (2008) took a different approach with MAST: mechanics, aesthetics, story and technology. Mechanics are once again the rules of play but here aesthetics are the appearance of the game. The story behind the game is then added as well as the technology used for play. In the MAST model the four elements of mechanic, aesthetic, story and technology are equally important and each one has an influence on the others (Schell, 2008). The 11 components of MAPS-AIM operate the same way. Each is equally important and changing one will impact the others. Getting each component right interactively creates the opportunity for a game to be greater than the sum of its parts.

The Eleven Components of Gamified Experiences

Six key components become immediately apparent from MDA, MAST and Costikyan's (2002) definition of games, and by extension gamified experiences: mechanic, aesthetic, poetic, technologic, dynamic and hedonic-eudaimonic. Looking at the game design literature, game-related research and actual games, five more components become apparent: schematic, referent, audience, intention and metric. The 11 components are divided into three layers in a visual model to emphasise their general and specific relationships (see Figure 3.1). The innermost is the content circle with mechanic (Mc), aesthetic (As), poetic (Po), schematic (Sc), technologic (Tc), referent (Rf) and dynamic (Dy). The next circle, experience, holds the hedonic-eudaimonic (HE). The connection circle has audience (Au), intention (In) and metric (Mt) and encircles the rest. This is the MAPS-AIM model of game components.

Making the model practical requires an example that can be carried through the description of the components. The tourism game that is proposed here is a modified treasure hunt in Abu Dhabi, UAE. This game can

Figure 3.2 Urban Excursion: Abu Dhabi: mobile app version

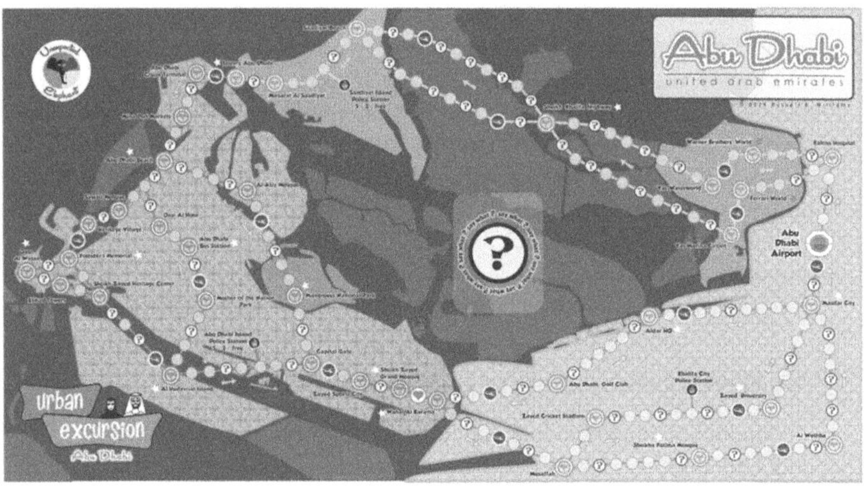

Figure 3.3 Urban Excursion: Abu Dhabi: board game version

be used promotionally by the city and Etihad Airways and is meant to engage visitors at a deeper level than would otherwise be experienced. Visitors to Abu Dhabi primarily imagine the place as a sun, sea, sand and shopping destination while Emiratis imagine their city as an expression of their traditions, history and aspirations (Slak-Valek & Williams, 2018). This game will send visitors on a race through important places in the city on the basis of self-selected themes, including identity, history, architecture, action, symbol and monument. Potential visitors could get a taste of the destination and the game on Etihad and Abu Dhabi websites or through play on the boardgame version (Figures 3.2 and 3.3). They could enjoy it further through the in-flight entertainment system on board inbound Etihad flights. Upon landing they would be invited to download the game app to their phone using the free airport Wi-Fi. They would be able to play the game at any point in their Abu Dhabi stay. This mobile location-based game, Urban Excursion: Abu Dhabi ©, will be the ongoing example used in this chapter.

The content circle

Mechanic (Mc)

The mechanic makes a game, or gamified experience, and completely separates it from other physical and mediated experiences. Without a mechanic there is no gamification. The mechanic includes three things: the goal of a game, the paradigm of play and the rules that place limits on player decisions and actions in pursuit of the goal.

The most important goal is winning the game. A winning state can be making the most money, having an empty hand, visiting the most attractions, becoming a hero, collecting the most Pokémon or simply surviving. The goal gives meaning to all of the objects, decisions and actions within the game (Burgun, 2015). Applying this, the goal in Urban Excursion: Abu Dhabi (UE:AD) is to visit more points-of-interest than opponents to win an individualised experience while in the city. Scoring the highest points within a specified three-day period could also win a significant prize for the return flight. There should also be sub-goals, so that players do not have to wait until the end to feel some sense of achievement and progression (Schell, 2008). Clearing levels, getting badges or earning upgrades of skills, knowledge or equipment are all forms of sub-goals. Goals should be concrete, achievable and rewarding (Schell, 2008). Concrete goals are easy to understand. Achievable goals are possible for players. Rewarding goals come with a sense of accomplishment, whether it is simply the act of completion or the importance of a prize within the game.

The paradigm for play is also part of the mechanic. There are three paradigms for any game: choice, chance and skill (Bateman, 2009b). Each paradigm can exist in a game alone or in combination. Combinations tend to be more interesting. Games of chance alone are essentially gambling where players are equal and winning is in the hands of fate (Caillois, 1961). Games of skill are for those with particular abilities in what is required to play. This can be sports, including e-sports, or athletic contests (Schell, 2008) where anyone can play but more skilled players win. While there are pure games of chance there is never a completely pure game of skill. Games of choice are cerebral or strategic with the pure examples of Chess and Go. Offering more than one paradigm of play within a gamified experience will do a better job of sustaining player interest (Lazzaro, 2009). Games can combine paradigms of play and provide a level playing field that rarely exists in reality. UE:AD combines all three paradigms, choice, chance and skill, in each of its different forms.

Figure 3.4 Urban Excursion: Abu Dhabi: question space

Rules establish what a player can and cannot do and the consequences of actions (Lindley, 2002). Rules are the agreed upon constraints for the duration of the game (Adams, 2010). They provide the pathway to the goal. When a game has more rules it is more complex (Bates, 2009b). Rules can be discovered in play or described in detail. In a professionalised sporting event, a referee polices the rules while in a match between friends the players police the rules. When programmed intelligence is involved, including mobile and augmented reality games, the programming polices the rules. 'By offloading the dull work of rules enforcement onto the computer, games can reach depths of complexity, subtlety, and richness that are not possible any other way' (Schell, 2008: 147). The central rule in UE:AD is that players cannot claim a location and move on to the next without answering a question about that location. The answer will be found by exploring the place, interacting with people and responding to hints if necessary. This rule will be enforced as a part of the app in the mobile form of the game (see Figure 3.4).

Operational rules are the kind found in rulebooks and instruction sheets. These are the 'rules of play' (Salen & Zimmerman, 2004). Some argue that the operational rules should be exact and unambiguous while others argue that some ambiguity in the specific rules can create higher levels of player engagement. There needs to be enough specificity that the game can be played with little effort, but enough latitude to make players a part of an emergent experience each time they play. Implicit rules are unwritten and vary from one context to another. These are the rules for conduct during play and are provided by the culture in which the game is played and the specific group with whom the game is being played. A game should have enough latitude that these 'house rules' can evolve from play. The operational rules of UE:AD are that the player(s) choose the locations to play (see Figure 3.4). They are then given GPS coordinates for an attraction which they must then visit. Upon arriving at the location they activate the question and it must be answered before they win the location and are given coordinates for the next location. The implicit rules are how they move from place to place and how they answer the site-specific question. Players can travel in any way they choose. Answering questions, they can explore the space, interact with other people at the site, visitors and locals, or use some other means to gather information.

Aesthetic (As)

Aesthetic is the point of connection between the internal aspects of the game and the players. Every game must take on a form that makes it usable for players. The aesthetic applies to any mediated presentation or device within a gamified experience. There are three key elements in aesthetic: the look, the interface and the feel. The look is the visual appearance of every aspect of a game. The interface is how players control their representatives in the game world and the ways in which they are able to interact within the world. The feel involves player point-of-view (POV), haptic elements and the audio environment.

The look is the most obvious part of a game. Poor writing and bad programming in a computer-enabled game may take some time to become apparent but bad visuals are obvious immediately (Adams, 2010). Strong visuals draw players into a game. They can strengthen the reality of the game world, or the connection between the game and the real world. Beautifully designed objects, environments and images can increase the value of the experience, increasing positive feelings for the game and enable players to overlook other weaknesses because of its beauty (Norman, 2004; Schell, 2008). A key element of look is the spectrum from realism to abstraction in the visuals. Jarvinen (2002) identifies three steps on that continuum: photorealism, caricaturism and abstractionistic imagery. Photorealism is the look of a live action film, where viewers imagine that what they see is a depiction of reality. Caricaturism is what is seen in an animated film where there is a range of styles from line drawings to near photorealistic work that is three-dimensional. Abstractionistic imagery in a game reduces the visuals to simple shapes, lines, icons and symbols with little connection to any known reality. A deck of cards or the stones and board used in Go are examples. UE:AD will include elements that are photorealistic combined with maps and icons that are on the caricaturistic side of abstraction (see Figure 3.5). It involves the connection between the photorealism of the place through live or stored images with the abstraction of a map and the comic feel of icons.

The interface includes the way players interact with the game world and the device used for that interaction. The device used to play physically-situated games can be a part of the human body or a direct extension of the player's body. When entering digital realms the device must enable a player to control objects and actions in virtual space. A controller

Figure 3.5 Urban Excursion: Abu Dhabi: aesthetic look of mobile app

providing an immediate understanding of its function for the user and its effect on the object that is being manipulated is thought to be the most natural and to have the most effective mapping (Norman, 1988). Yet, when players have learned to use a less natural controller, their creation of a mental map for using the device is more important than the accuracy of the conceptual map used in design (Williams, 2018). The screen side of interface comes down to either diegetic or non-diegetic objects (Mitchell, 2012). Diegetic objects are a direct part of the environment. Interaction with these does not take the player out of the game world. Non-diegetic objects are those that are not a direct part of the environment. These include control panels, heads-up displays and health meters. Augmented reality interfaces place non-diegetic objects over reality and/or enhance reality with computer-generated diegetic objects. The keys for all on-screen interface objects is that they are appropriate to the environment and the action, responsive to the actions of the player and changes in the game, and transparent. Transparency enables a player to focus on the content and play, rather than the interface. UE:AD will include augmentations to the realities of the location and diegetic elements that can be manipulated through the touch screen of a mobile phone.

Feel includes player POV, haptic elements and sound. Player POV includes first person, third-person immanent and third-person transcendent. First person is through the eyes of a character; the player becomes a character in the game. Third-person immanent is viewing an experience from near the perspective of the on-screen avatar who can also be seen in the field of view. The character is seen separately from the player but they are having the same visual and visceral experience. Third-person transcendent is from beyond what is humanly possible or looking down on the action from above. It is easier for a player to get involved in a game when they are playing in first person POV (Lecky-Thompson, 2008). Each one of these points of view will have a different psychological effect on the player (Moore, 2011). The POV in UE:AD will be either first person or third-person transcendant. Third-person transcendant will be the maps that direct play and first person will be in the views of the locations and the augmentations of that reality.

Haptic elements are literally about feel. These involve any attempt to give a sensation of touch through a control device. These include controllers that vibrate in concert with action on the screen (Adams, 2010) and devices that give 'force feedback' in an attempt to replicate real-world physics. The haptic is also important in more traditional game forms. Playing cards have a very specific texture and stiffness that make them feel right and makes it possible to shuffle them or perform slight-of-hand. Dice have a certain weight and balance and game tokens a particular feel between thumb and finger that can have an impact on enjoyment of a game.

Sound is the third part of feel. Well-executed audio creates a full and complete gameplay experience (Mitchell, 2012). Diegetic sounds are those that actually come from the environment portrayed. These are sound effects and ambient sound. Factual or fictional diegetic sound helps create a complete picture of the game world, even when play occurs in the physical world. Non-diegetic sounds are those that are not specific to the environment. Non-diegetic sounds are primarily musical. Music can be a part of an environment but generally it is not. Both diegetic and non-diegetic sound can be used to communicate the invisible or what's coming next. The most commonly used audio in games is sound effects. Good sound effects are based on the psychological expectations of players and not the actual recording of the sound in question (Adams, 2010). Thinking a sound is accurate is more important than its actual source. Sounds should reinforce the player's point of view, what is visible and what is not (Jarvinen, 2002; Lecky-Thompson, 2008).

Music works very well to create mood and establish intensity. When music is done right, a person can understand what is going on with their eyes closed (Lecky-Thompson, 2008). Studies have shown that listening to music can release dopamine in the brain giving a very pleasurable experience (Despain, 2013) and this can be connected to game play. The use of pop, rock and hip-hop music that has no direct connection to a game (Jarvinen, 2002) is a phenomenon that can also effect the emotional state of players (see Figure 3.6).

Figure 3.6 Urban Excursion: Abu Dhabi: sound elements of locations

Poetic

The Poetic is the story in the game. It is the connective tissue between other game components and an opportunity for increased player engagement. The Poetic 'provides the players/protagonists with goals, conflict and an element of uncertainty. From these narrative cornerstones arises dramatic tension, an experience that is sought and encountered in computer games and as well as fiction films' (Rusch, 2008: 24). The advantages of a story in a game are that it gives the other components structure and meaning and it provides a beginning, middle and end that Aristotle would appreciate (Moore, 2011).

Academics have looked at the impact of story in games and found that a Poetic has primarily positive effects. Schneider *et al.* (2004) found the players in first person shooter games identified more with characters and experienced a stronger sense of presence when there was a story attached to the game. Pre-game stories created the opportunity for quicker understanding of a new game world and the relationships within it (Park *et al.*, 2010). This can hold true when the 'game world' exists in the physical space of a destination or attraction or the imagined space of fantasy or imagination. Balakrishnan and Sundar (2011) found that strong narrative increased enjoyment. 'Narrative experience in games can take two forms. There is the embedded narrative that consists of the pre-scripted moments and structures that are relatively fixed in the game. But the true strength of the games on the narrative level lies on the emergent narrative, the narrative that arises during play often in unexpected ways, as a result of the individual moment-to-moment game-play' (Rusch, 2008: 24). A Poetic can be fixed 'with prewritten narrative chunks' or be emergent, 'arising out of the core mechanics' (Adams, 2010). Stories are about the choices characters make and may hinge on a key choice made by the protagonist (Crawford, 2013). Players become protagonists in the game Poetic as they make the choices based on the story and the Mechanic.

A good Poetic increases interest in a game and makes it easier to understand the Mechanic. 'Not only does the narrative give extra meaning to the rules, but the rules help to perceive the narrative' (Salen & Zimmerman, 2004: 387). A story can give meaning to otherwise obscure Mechanics (Burgun, 2015; Schell, 2008). It provides context, is the key to pretending within the magic circle, and provides dramatic motivation toward a goal that might otherwise be completely abstract (Adams, 2010). If the Poetic is too complex players can spend cognitive capacity that could serve them better in other parts of the game. People have a limited amount of cognitive capacity and they must prioritize according to their interests and goals (Lang, 2000). Lindley (2002) describes a 'narrative gestalt' that is developed to manage the narrative elements and structure in a game. There must be a balance or an acceptance that gameplay or narrative will dominate the entire experience of play. Narrative and gameplay can also alternate in importance for the player. Applying this, the poetic in UE:AD will be largely emergent. It will be the story of the visitors adventure in Abu Dhabi. It will be their 'extraordinary story.' There will be connections made between the ordering of locations and a story that can be told on that basis, but that story will be understood and interpreted by the individual player.

Schematic

The Schematic is a representation of the relationships between objects and players in the game-space. 'Games need to have a strong narrative to make sense, but they also need to have a strong and structured environment to reinforce that. Whether real locations or abstract patterns, formal constraints and ordering devices for levels can be drawn from different sources' (Brown, 2002). Every game has a space in which people play, even when the game is not location based. The Schematic is essentially the game map. It can be generated *ex nihilo*, be a product of synthetic design, or be derived from the abstraction of a process, system, or environment from the physical world. 'Schematics are the first step of a design proposal, much like pre-production' (Brown, 2002). 'When you can think of your space in these pure abstract terms, it helps you let go of assumptions about the real world, and it lets you focus on the kinds of gameplay interactions you would like to see' (Schell, 2008: 135).

The Schematic is the representation of relationships rather than physical reality. The represented relationships can be spatial, social, cultural or perceptual. An enduring image of a schematic representation is the 1933 underground map for London. Designed by Harry Beck, the 1933 map tells the user about the relationship between the stations on the

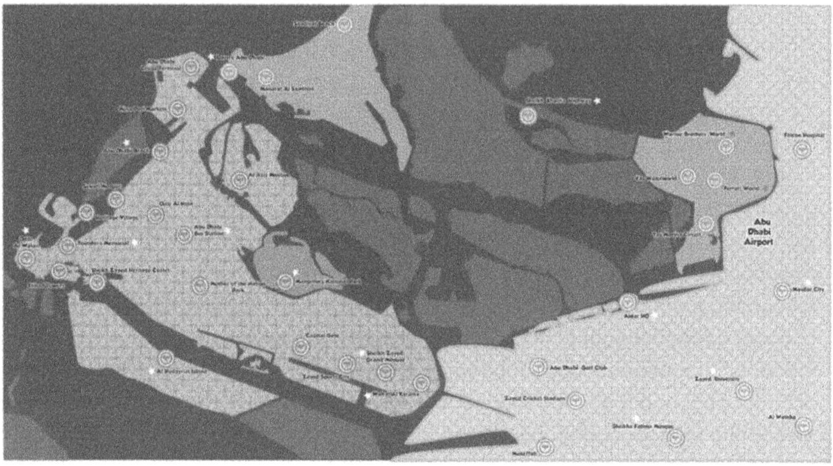

Figure 3.7 Urban Excursion: Abu Dhabi: Locations in City on Game Map

underground lines rather than the actual topography of London. The map will not help a person on a walk through London but it will help a person use the underground to get to their desired destination. 'We need to strip away all visuals, all aesthetics, and simply look at the abstract construction of a game's space' and the relationship between places in the game (Schell, 2008: 130). 'It is possible to sample a location's organization, and re-use it at a different scale in another context' (Brown, 2002).

Games are essentially abstractions of reality (Moore, 2011). The only way to include every element and variable found in the real world is to play the game in the real world. The Schematic reveals the true structure of the game-space in what can be reasonably called a map. Some games are essentially played on the map itself as an absolute abstraction of reality (Jarvinen, 2002). From this point it is possible to determine whether the game should be on rails or an open world. A game on rails has a fixed linear pathway for players to follow. 'These experiences excel at focusing player attention on specific tasks and key moments' (Despain, 2013: 160). An open world allows players to choose their paths through the game. There are always limitations to the openness of the world for the sake of progression and scale, but a pure open world allows the individual to play any way they choose within the given space.

The game map in UE:AD is the map of the city (see Figure 3.7). It is a gamified way to traverse the city and its points of interest. A visitor could certainly just use a map and guidebook to travel around the city and discover interesting locations, but this gamified version could prove to be more engaging (see Figure 3.8). The relationships in the schematic come from the physical relationships in the Abu Dhabi environment (see Figures 3.9 and 3.10).

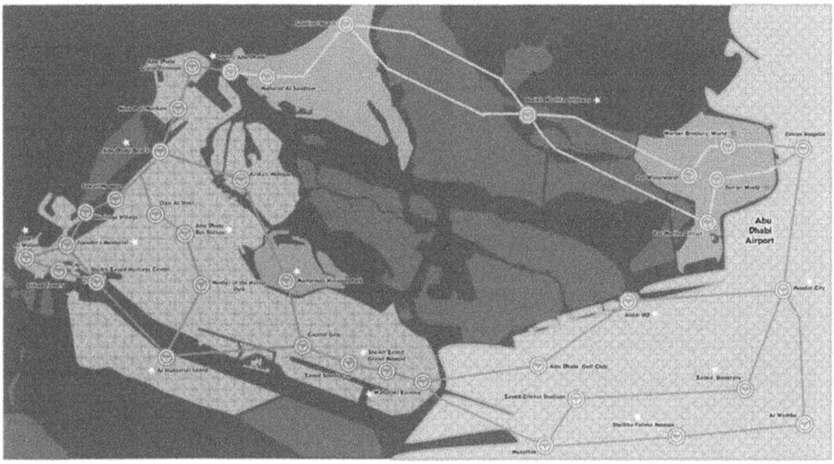

Figure 3.8 Urban Excursion: Abu Dhabi: Linking Locations on Game Map

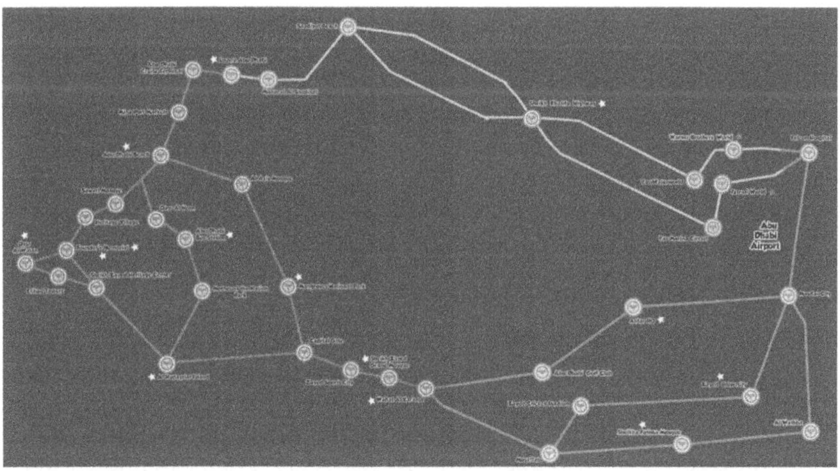

Figure 3.9 Urban Excursion: Abu Dhabi: Isolated Schematic of Game Map

Technologic

The Technologic of a game is the technology necessary for people to play. 'The technology you choose for your game enables it to do certain things and prohibits it from doing other things' (Schell, 2008: 42). This includes technology in the physical and mediated realms. Costikyan (2002) calls games 'amazingly plastic' as they have adapted to every available technology from 'the Neolithic to the high tech' (9). Even games that appear to be without devices and played physically often require some kind of technology, a ball, chalk, a basket, or markers. A few childhood

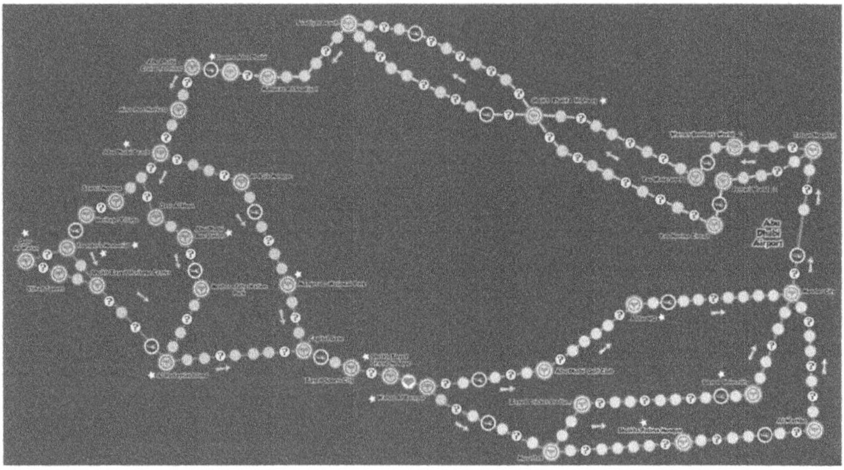

Figure 3.10 Urban Excursion: Abu Dhabi: Board Game Schematic of Game Map

games do not require any Technologic, while games played with boards, cards, stones, dice, computers, game systems and mobile devices involve increasingly complex forms of technology.

The Technologic used enhances the possibilities for play and extends the possibilities for human interaction. The most frequent Technologic is some form of media. There are 10 dimensions of media that enable a wide variety of gameplay experiences (Williams, 2014): (1) Line; (2) Plane; (3) Space; (4) Sequence; (5) Synergy; (6) Time; (7) Control; (8) Intelligence; (9) Presence: (10) Distance. Next, we will discuss these dimensions in details.

Line primarily represents potential but a simple line can divide two players and be the basis for a game. Plane is the foundation for graphic media with the existence of height and width and the possibility of using these to create the appearance of depth. This is the realm of images and maps. Space consists of actual height, width and depth. Discrete objects also possess these attributes. These are real objects that can possess internal as well as external space. When graphic or discrete media are put into sequence then textual media become possible. The sequence of objects creates meaning. Synergy is the combination of graphic, discrete and textual media to create meaning. This is the original form of multimedia as media with different dimensions function together. Time is the basis of temporal media, i.e. music and motion pictures. This is a source of tension in games as fixed temporal sequences exclude choice, chance and strategy. Temporal media can display games and enhance a game experience but not be a complete game platform. Control is the seventh media dimension. Interactive media offering this dimension give control to media consumers and they can become players. This is as simple as a pause button or a vending machine. This returns control to the player after the imposition of time, and can be executed without the intervention of digital technology. Intelligence is the dimension that makes interesting gameplay possible in the digital realm. Control is not enough. The analogical media made possible with this dimension can appear to possess the qualities of anything that exists in the physical world. Programmed intelligence makes this possible. Intelligence also makes it possible to enforce game rules, offer virtual opponents and non-playing characters, and respond according to the skills of a player. Presence gives a person the sense that they are a part of the content and not operating outside of it. This is virtual media and is strengthened by the first person POV and/or a personalized avatar. Augmented Reality relies heavily on this dimension. When presence is fully realized the player is completely engaged in a virtual reality. Distance can be physical, sociological, cultural, religious and/or political. In connected media these various distances between people are minimized. A telephone mediates between people reducing physical distance. An MMORPG can reduce all distances between people with the added motivation of the game.

Playing a game in the physical world can involve the use of any or all of these types of media. Direct human interaction is the simplest and yet the most complex form of interaction. It can exist without mediated objects or it can incorporate some or all dimensions of mediated objects. A tourism game that engages players in the physical space of an attraction or destination can be amplified by the sophisticated use of these various media elements. UE:AD played live is that kind of game as it uses all dimensions and forms of media through the technological opportunities offered by a mobile device.

Referent

The Referent is the process, system or environment (PSE) that serves as the conceptual basis for a game. While games often appear to be pure abstraction, 'All play in some sense represents something from the non-play universe. We often confuse this metaphorical aspect of play as simulation. Play is not necessarily a simulation of anything in particular, but it does generate mental associations with real-world issues' (Crawford, 2003: 14). Snakes and Ladders has a referent from India, 'The game was designed to teach children about good and evil. If the player landed on a good square, they could continue to move up in the game. Landing on an evil square caused the player to fall toward the bottom' (Mitchell, 2012: 11).

There are five aspects of a PSE that will have an impact on the game: the function(s) of objects/people in the PSE; the core relationships between objects/people in the PSE; the goals of objects/people in the PSE; the barriers to the achievement of goals; and the affordances for the achievement of goals. Barriers are the obstacles to action. Affordances are the actionable properties between the world and an actor (Norman, 1999). Identifying these attributes is key in the process of game design and gamification. It is accomplished by systematically removing objects and representations which are part of the reality but do not inform an understanding of the PSE. Abstraction leads to the conceptual model of the PSE. The other game components can then grow out of this conceptual model. The Referent requires opportunities for planning, intention and consequence on the part of a player (Church, 1999). These can be found in answers to three questions. What features of the PSE make it playable? Where are opportunities for scoring? How is effort and reward balanced in the PSE? 'The point is that you don't need to perfectly replicate real experiences to make a good game. What you need to do is to capture the essence of those experiences for your game' (Schell, 2008: 20). The result of this process always delivers compromises on reality in service of playability and the subjective experience of players (Adams, 2010). The gamification of real-world leisure and tourism also benefits from this process of abstraction and establishment of the Referent. City-based tourism games can use the city itself or some aspect of history as the referent. Larger spaces such as a region or country and smaller spaces such as an attraction – museum, amusement park or urban space – can also serve as physical points of reference. A game in a wilderness area and an urban area can have preservation for future generations as a referent. The possibilities are as broad as tourist motivations and may create the opportunity for deeper engagement in the destination or attraction. The city and emirate of Abu Dhabi is the referent of UE:AD.

Dynamic

The Dynamic is what players will do in a game. This is perhaps confusingly called game genre because it is not the same as genre in other media. The Dynamic is the categories that represent particular game challenges and have no relative consideration of setting or content (Adams, 2010). There are 10 general categories of game Dynamics: Casual; Deliberate; Serious; Social; Solitary; Open World; On Rails; Board Games; Card Games; Physical Games.

Casual games are easy to learn and don't require a great deal of time to play (Mitchell, 2012). Classic addictive mobile games fit into this category and can involve different specific dynamics. Deliberate games require long play sessions and player commitment. They can also have a significant learning curve. A deliberate game requires conscious effort in terms of time and learning. Serious games are meant to teach, persuade or convey important information. They are more than amusement. Social games require people to play in groups. Solitary games only require one person to play. Open-World games allow players to choose their pathway through the game, which mission they will face and the frequency of pursuing missions. On-Rails games guide the player through the game, developing the skills and knowledge required for each task and revealing the game world in a controlled way.

Board games and card games are defined by their traditional form regardless of their current Technologic. These are the games played before the advent of computers but continue to be rich sources of gameplay experience. Physical games involve engagement of the human body. Physical games traditionally relate to playground and sportsground but now relate to kinisthetic controllers, location-based gaming and augmented reality as well. A physical game can't be played sitting in a chair. UE:AD is an open-world physical-game and also an On-Rails boardgame.

Specific dynamics are detailed descriptions of player activities. An alphabetical, but not exhaustive, list includes Adventure, Action, Economic; Fighting, First Person Shooter (FPS), Platform, Puzzle, Rhythm; Role Playing (RPG); Simulations, Sports, Strategy, Survival and Trivia.

Adventure games allow players to move at their own pace through the game world to discover its treasures, solve puzzles and experience the Poetic (Mitchell, 2012). This can be a particularly fruitful dynamic for tourism-centered and location-based games. There can be combat in an adventure game but when it is emphasized it becomes an Action game (Despain, 2013). Economic games focus on the costs and benefits of various choices with a focus on finance. Fighting games are one-on-one, one on many or many on many battles in an arena-like setting, involving fighting styles from boxing to martial arts and a wide range of weapons and abilities. An FPS is played in first-person POV with the weapon visible to the player. Players must kill enemies and destroy enemy resources while preserving their own. Ammunition management and stealth are key in these games (Adams, 2010).

Platform games require players to jump their avatar from one platform to another, overcoming differences in height, distance, obstacles and enemies. Platformers tend to be cartoonish with unrealistic physics (Adams, 2010), examples include *Mario* and *Sonic* and a large number of mobile games, including *Subway Surfer*. Location-based platform games present interesting opportunities for destination promotion. Puzzle games require the player to solve abstract object combinations, shooting angles and ricochets or realistic conundrums in the game world. These can be situated within real-world environments. Rhythm games have players follow rhythms and patterns in the game. This specifically includes dancing and instrument play.

RPGs have players become a character in a richly-fashioned game world and Poetic, and develop the strengths and weaknesses of characters over the course of play. Online versions involve large groups of people playing with some cooperating in teams of variously enabled characters and competing with others in ways dictated by the game.

Simulation games have at least four different forms which all mimic reality. Players in construction/management simulations build anything from a civilization to an amusement park and manage it toward a goal. Driving and flight simulators invite players to drive cars or fly planes in the particular simulacrum of the game. Life simulators make players the god of the game world creating every aspect of life in the world. Sports in the digital realm can be considered a simulation because it is important to make it as close to the reality as possible. Real-world sports are also games.

Strategy games are thoughtful experiences where players must overcome conflicts through strategic planning, tactics and sometimes logistics and economics (Mitchell, 2012; Adams, 2010). Survival games have one goal, to be alive at the end. Trivia games are based on players' knowledge, specifically and/or generally.

The list of exact actions that can be taken in a game is continuously expanding and evolving with changes in game-related technologies and approaches to gamification. UE:AD is an adventure game that includes puzzles and trivia.

The Experience Circle

Hedonic/Eudaimonic

The subjective experience in gameplay can be Hedonic, commonly referred to as fun, or it can be Eudaimonic, something a great deal more serious. 'Eudaimonia reflects... a greater interest in consuming entertainment that is associated with affective responses that reflect a blend of poignant and emotional reactions' (Oliver & Raney, 2011: 1001). 'Life is the struggle for survival and growth. There is no end to strife, not this side of the grave. A game without struggle is a game that's dead. We equate struggle and work and obstacles with pain, not pleasure. But it is absolutely true of games. We want games to challenge us. We want to work at them. They aren't any fun if they're too simple, too easy, if we zip through them and get to the end screen without being challenged' (Costikyan, 2002: 17). Games can offer more than hedonic entertainment and meeting different player needs for subjective experience is associated with different game features (Elson *et al.*, 2014). The subjective experience of games can be very similar to tourism. Enjoyment is not simply found in producing happy, smiling, faces.

LeBlanc (2004) identified eight kinds of 'fun' in relation to playing games. There are clear connections here to tourism fun. Sensation enjoyment involves game as pleasurable to the senses, primarily visual and aural. Fantasy involves games as make-believe in an imaginary world where a player can be anyone they choose. Narrative enjoyment of a game is in the drama that comes from the sequence of experienced events. Challenge is a necessary source of fun and comes from games as obstacle

Table 3.1 Three perspectives on hedonic and eudaimonic enjoyment

LeBlanc (2004)	Schell (2008)	Lazzaro (2009)
Sensation	Sensation	Hard Fun: Mastery
Fantasy	Fantasy	frustration
Narrative	Narrative	overcoming obstacles
Challenge	Challenge	relief
Fellowship	Fellowship	Easy Fun: Exploration
Discovery	Discovery	curiosity
Expression	Expression	surprise
Submission	Submission	wonder
	Anticipation	awe
	Schadenfreude	Serious Fun: Purpose
	Gift-Giving	excitement
	Humour	focus
	Pride of Accomplishment	relaxation
	Purification	People Fun: Social
	Surprise	gratitude
	Thrill	schadenfreude
	Triumph Over Adversity	amusement
	Wonder	admiration
		social bonding
		pride in helping

courses or problem-solving exercises. Fellowship enjoyment comes from a game as a social framework. The fun of Discovery comes from a game as uncharted territory, seeking and finding new and interesting things about the game world or the best way to play. Expression enjoyment involves game as a journey of self-discovery and a means to express self. The fun of Submission comes from leaving the real world to enter the game world and submit to the rules, goals and environment. These types of fun include a mix of Hedonic and Eudaimonic pleasures. Schell (2008) adds other types of fun to this list, including Anticipation, *Schadenfreude* – delight in the misfortunes of others, Gift-Giving, Humour, Pride of Accomplishment, Purification, Surprise, Thrill, Triumph over Adversity and Wonder. Table 3.1 summaries the above elements.

Lazzaro (2009) identifies four types of experiences players like about games and calls them fun. This fun is also both Hedonic and Eudaimonic. The opportunity to face a challenge and then master the required skills is Hard Fun. Easy Fun involves the inspiration that comes from sparking the imagination and the opportunity to simply explore. The opportunity to play with purpose and change oneself or the world is Serious Fun. People Fun comes from hanging out and playing with friends.

The progression of emotions in Hard Fun is frustration, *fiero*, the satisfaction of overcoming obstacles, and relief. The progression of emotions in Easy Fun is curiosity, surprise, wonder and awe. It's possible when players can go off-the-rails and explore. The progression of emotions in Serious Fun is excitement, focus and relaxation that come from repetition, rhythm and collection. People Fun is the product of social play. The emotional results of People Fun are gratitude, schadenfreude, amusement, admiration, *amiero* – social bonding – and *naches* – pride in the accomplishment of someone you have helped (Lazarro, 2009). There are similar experiences in tourism terms. Thus gamifying any aspect of tourism does not mean that it only involves positive emotions and outcomes. There is value to be had in struggle and the recognition of the struggles of others, whether we are speaking of games or tourism.

The Connection Circle

Audience

The intended Audience is the first thing to be considered when creating a gamified experience (Mitchell, 2012). 'You have to think about who would enjoy that experience' (Adams, 2010). It is imperative to know who will be playing a game because their identity is important for every component. 'Different people enjoy different media' (Lecky-Thompson, 2008: 28). 'Few games aimed at children opt for photorealism. Instead, their style is closer to the caricaturism of cartoons... the emotional and aesthetic experiences they provide can feel, at least in the minds of children, quite lively and real' (Jarvinen, 2002: 123). The definition of a primary audience for a game and a tourism offer is absolutely imperative.

Makkonen and Hokkanen (2013) studied the development of a game for a museum in Finland. The game was 'designed and developed for all user groups regardless of age and gender' (261). It was not surprising that the game was not viewed as valuable for the attraction and region. It is not advisable to design an experience for everyone. The audience needs to be well defined and understood. Battista *et al.* (2016) studied a geo-caching game, a treasure hunt using GPS, with rural 14 to 18-year-old high school students in western North Carolina. Their focus enabled them to produce significant results and discover psychological variables that played a part in participation. This is a well-defined audience playing an age- and interest-appropriate game.

The five key elements of Audience are demographics, psychographics, personality, cognitive development and motor skills. Demographics are the most basic way to identify the members of an Audience. They are easily quantifiable and include age, gender, education, income, residence, religion and ethnicity. Though packed with information, demographics can be easily misunderstood. Psychographics include lifestyle, attitude, values and preferences. These Audience measures cut across demographics and provide a richer picture of the targeted players.

Drilling deeply into attributes that can be shared by people Van den Berghe (2012) emphasizes the importance of the Big Five personality traits – openness, conscientiousness, extraversion, agreeableness and neuroticism. He posits that people with a dominant trait will have interest in specific kinds of experiences. Openness correlates with novelty and unexpected events. Conscientiousness connects with challenge at a high level. Extraversion corresponds with the stimulation that comes from social interaction. Agreeableness correlates with a desire for harmony and cooperative experience. Neuroticism, a tendency to experience negative emotions, connects positively with threat as people who score high on this trait seek out experiences that activate negative emotions. Understanding audience personality will deeply inform decisions in every other component of a game.

Cognitive development and physical skills are also important. Knowing what players are able to do is important before setting them a task. Players have differing competencies particularly when age is involved (Ibister, 2009). A player may not have the cognitive development necessary to understand a game mechanic and its expression in play. Cognition improves with age through education and experience but it can diminish in later years. Players may not have the motor skills required for some activities. 'Keep in mind that some players may be unable to master a certain motor control issue due to age or accessibility' (Despain, 2013: 142). Motor skills develop and then diminish on a U-shaped curve over a lifetime. Thompson, Blair and Henrey (2014) found that slowing response times in gameplay begins at 24-years-old and game-specific skill does not impact this decline.

It is as important to identify the entry behaviors, abilities and knowledge players must possess to play a game, grasp its meaning or enjoy the experience as it is to profile the audience with demographics, psychographics and personality. Identifying the target audience is an obvious component of game, and gamification, design in tourism but it seems to be easily overlooked.

The primary audience for UE:AD crosses genders and ranges in age from 14 to 25. Teen tourists will be travelling with parents and can push them to be a part of the game experience. Players will be in secondary school or university and enjoy learning about the places they visit. They will not be completely happy with laying on the beach and want to have interesting and informative content for their tweets and posts on Instagram. They will also have an interest in direct human interaction at the destination. Their key personality traits will be openness and extraversion. They will be interested in making their own choices about their trip and appreciate that aspect of the game. The secondary audience will be in the 25–34 age-range. They will possess at least a university degree and will share the personality and psychographic attributes of the primary target group.

Intention

Games communicate ideas, attitudes, behaviors, information, and more. The Intention of the game is what the creators intend to communicate to the audience. 'To affect an audience in a meaningful way requires more than just clever play. It demands considerable resolve' (Despain, 2013: 78). Adams (2010) said the 'duty' of a game is to entertain and all else is secondary. But there is more to intention than entertaining. 'When the designer takes the time to identify the goal of the entire experience, the path to great gameplay starts to appear. Use the designing experience as an opportunity to challenge broader thinking, to engage unique themes, and to delight and surprise the gamer' (Despain, 2013: 78). Intention includes the communication goal of the game and the underlying theme.

There are seven communication goals in the creation of media content and games (Williams, 2014). These Intentions are to inform, connect, attract, record, persuade, entertain, and express. Informing is providing information to players. It can be understood as journalism, education or training. Connecting is providing the opportunity for players to interact with others and to possibly build a community. Attracting is simply getting people to take notice. Recording is capturing a moment or event for future reference and remembrance. This is not often an intention for making a game but it could be. Persuading is getting players to agree with a particular point of view, attitude or behavior, whether it is commercial, social, political or religious. Entertaining is providing players with diversion, relaxation, recreation, or amusement. If it is not the primary intention of the game, it certainly needs to be the secondary one. Expressing is the realm of culture, or art, where the game maker uses the medium to touch players and make them feel something, communicate a culture or simply make something beautiful.

Most games have at least two intentions, a primary and a secondary, and there should not be more. Trying to do everything, just like trying to reach everyone, will prove unsuccessful. Serious games can have the primary intention to inform, but they must also entertain to engage the audience. However, 'the accuracy of the process or effect being simulated for training is of primary importance. In addition to the accuracy of what's being taught, the serious game must also be concerned with whether – and what – the game is actually teaching the player. If the player learns to beat the game but cannot usefully apply what he is learned in the real world, then the serious game has failed its mission' (Michael & Chen, 2006: 43).

The two intentions of UE:AD are to inform the players about Abu Dhabi and to connect the players with other tourists and city residents.

Once a communication goal is established then it becomes important to devise a theme. 'Theme is the central idea your design will convey. It's the one big concept that drives all of the work you do. It's the part of your design that adds substance and makes people see value beyond the fun they'll have or the skills they'll learn' (Despain, 2013: 120). Theme can be understood as Occam's Razor in the design process, it is the measure that determines what is too much and what is too little. 'The primary benefit of basing your design around a single theme is that all of the elements of your game will reinforce one another, since they will all be working toward a common goal… [when] deciding if something belongs in your game or not: If it reinforces the theme, it stays, but if it doesn't, it goes' (Schell, 2008: 49). 'For theme to be the most useful, it needs to be as specific and targeted as possible. As an example, "war" is simply too broad to be a helpful theme. Is the game about the horrors of war? Or maybe about honoring the soldiers who fight? In looking at just those two thematic ideas, some very important differences dictate significantly different implementations in design' (Despain, 2013: 120). The theme of UE:AD is that there is more to Abu Dhabi than sea, sand, shopping and skyscrapers. The tagline for Abu Dhabi Tourism is 'Your Extraordinary Story' and the game will help visitors write their own story.

Themes are deeply connected to communication goals. Both should also be manifested clearly in the Poetic. 'As theme relates to narrative, in particular, the main character's story arc (the change in a character from the beginning of the story to the end) usually acts out the theme most clearly. In almost every case, whatever the protagonist of the story learns or experiences is what the player will learn or consider' (Despain, 2013: 120). Despain calls theme the backbone of game design. 'It will be a guide throughout the creative process and offer the player meaning and value beyond the fun of gameplay' (2013: 120).

Metric

The Metric is the measure of success in a game (Williams, 2014) or a gamified experience. Whether or not an intention was achieved should be a metric. There are ways to measure this type of success and it is important to do. The metric that gets the most attention is financial. Games that make more money are considered more successful regardless of other measures of success. Not all games make money the same way. Games can now operate with a number of business models that enable the game to be successful financially. The nature of the business model, the metric, has an impact on other components.

The traditional way to make money with a game is to sell individual games. The number of units sold connects with a level of revenue and the game makes money after the costs of production, distribution and marketing have been met. This metric has a limited effect on the other components except that the game must find its audience and those people need to be willing to pay for the experience offered. A very old business model for games was the coin-drop in an arcade. The coin-drop metric depended on the popularity of the game to get it into the arcade and then to get people to play it one coin at a time. This metric had a direct impact on other components as the game needed to work in small segments but players needed to be confident that their investment would eventually result in success.

An increasingly important way for games to make money is through in-game advertising. Technologics make this easier than others. There are two types of advertising that can occur in-game, interruptive and integrated (Williams, 2020). Interruptive advertising is traditional advertising. It is self-contained content and has no direct connection to the content in which it is placed. It disrupts the flow of that content. A game might get enough attention through downloads, views, likes, shares and plays that advertisers think it can deliver them an audience that may buy what they are selling. They are then willing to put interruptive advertisements in and around the game and pay for that opportunity. Integrative advertising fits within the content where it appears, it is part of the story, the action or the environment. This can be product placement. This is where an advertiser thinks a game will deliver an audience that interests them so they pay in advance for their seamless presence in the game. Integrated advertising is a part of the play experience and not a distraction from it. The integrative and interruptive advertising Metrics can have an impact on other game components as they require an opportunity for the advertisements to exist in the game world. Locating games in specific destinations and attractions in the same way films are shot in real-word locations is a great opportunity to promote those destinations and attractions, particularly when the audience for the game fits with the targeted audience for the tourism location.

Another important way for games and gamified experiences to make money is through the *freemium* model and in-game purchases. With the freemium model the player gets the game for free, the core product, but must pay a premium for additional features, products, goods, functionality, cheats, mods, abilities, tools, weapons, characters and anything else that can add to the experience of play. The freemium metric has a great deal of impact on other components, as they must serve the idea that players need to be encouraged to spend money on upgrades after they have gotten the game for free.

There are other metrics that have arisen around games and that will arise in the future. Some of these will have direct impact on game components and others will not. For games outside of the mainstream and with very niche appeal, crowd sourcing has become a legitimate means for game development and distribution. The crowd-sourcing metric is mostly about finding an audience but can impact components if there need to be premiums offered for higher contributions toward game development and distribution.

The most important metric for UE:AD is the number of people playing the game and the time they spend on the app while in the Emirate. Associated with that is the increase in visitor numbers at the locations included in the game. The downloading of the game app is a metric that is often reported, but it is meaningless if people don't log-on and play. In this case this means going to a wider variety of attractions.

Conclusion

Carl Jung, psychiatrist and founder of analytical psychology, had a deep love for games. He said, 'One of the most difficult tasks men can perform... is the invention of good games, and it cannot be done by men out of touch with their instinctive selves' (Van der Post, 2010: 41). MAPS-AIM is an attempt to systematize some of the invention and make it possible for game designers and gamified experience creators to stay in touch with their instinctive selves. This will open opportunities for creativity and innovation in games and tourism and put analysis of that work on the same foundation. This is extremely important as the gamification of tourism experiences and promotion continues to expand.

Harmonizing the work of LeBlanc (2004) and Schell (2008) and probing deeply into the game literature it is evident that there are at least

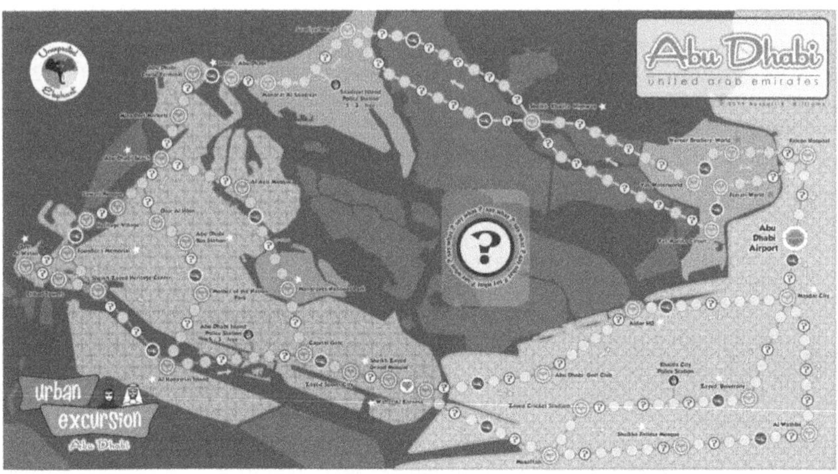

Figure 3.11 Urban Excursion: Abu Dhabi: Board Game

11 fundamental game components. The inner Content circle includes the Mechanic with goals, paradigm and rules, the Aesthetic with look, interface and feel, the Poetic with fixed and emergent stories, the Schematic, the Technologic with 11 dimensions of interaction, the Referent and the general and specific categories of the Dynamic. The Experience circle includes only the subjective experience of the Hedonic-Eudaimonic. Audience with demographics, psychographics, personality, cognitive development and motor skills; Intention with communication goals and theme and Metric with business models and various measures of success are in the outer Connection circle. These components are interrelated and as one changes the others must be adjusted in varying degrees.

An 'audience isn't playing the game merely because the designer is innovative in the way they approach physics-based puzzle design or the story's plot twist; it's a combination of all elements... [and] it needs to become greater than the sum of its parts' (Despain, 2013: 78). This is true for all games and gamified experiences regardless of application and intention. A game that is related to tourism must compete with all other gaming experiences available to the target audience (see Figure 3.11). Tourism games and gamified experiences must take advantage of every possible component to achieve excellence, attract and engage players who are, have been or will become, visitors to the destination or attraction.

References

Adams, E. (2010) *Fundamentals of Game Design* (3rd edn). Berkeley, CA: New Riders.

Balakrishnan, B. and Sundar, S.S. (2011) Where am I? How can I get there? Impact of navigability and narrative transportation on spatial presence. *Human–Computer Interaction* 26, 161–204.

Bartle, R.A. (2009) Understand the limits of theory. In C. Bateman (ed.) *Beyond Game Design: Nine Steps Toward Creating Better Videogames* (pp. 117–133). Boston, MA: Charles River Media.

Bateman, C. (2009a) Include players with different skills. In C. Bateman (ed.) *Beyond Game Design: Nine Steps Toward Creating Better Videogames* (pp. 189–212). Boston, MA: Charles River Media.

Bateman, C. (2009b) Understand patterns of play. In C. Bateman (ed.) *Beyond Game Design: Nine Steps Toward Creating Better Videogames* (pp. 61–116). Boston, MA: Charles River Media.

Battista, R.A., West, S.T., Mackenzie, S.H. and Son, J. (2016) Is this exercise? No, it's geocaching! Exploring factors related to aspects of geocaching participation. *Journal of Parks and Recreation Administration* 34 (2), 30–48.

Brown, D. (2002) GDC 2002: Manhattan as Muse: New York City as a conceptual tool. *Gamasutra*. Available at: http://www.gamasutra.com/features/20020417/brown_01.htm(accessed July 2016).

Burgun, K. (2015) *Clockwork Game Design*. Burlington, MA: Focal Press.

Caillois, R. (1961) *Man, Play and Games* M. Barash (trans). New York: Simon and Schuster.

Church, D. (1999) Formal Abstract Design Tools. *Gamasutra*. See www.gamasutra.com/features/19990716/design_tools_01.htm (accessed July 2016).

Costikyan, G. (2002) I have no words and I must design: Toward a critical vocabulary for games. In F. Mayra (ed.) Proceedings of the Computer Games and Digital Cultures Conference (pp. 9–34). 6–8 June 2002, Tampere, Finland.

Crawford, C. (2013) *Chris Crawford On Interactive Storytelling* (2nd edn). Berkeley, CA: New Riders.

Crawford, C. (2003) *Chris Crawford on Game Design*. Berkeley, CA: New Riders.

Despain, W. (2013) *100 Principles of Game Design*. Berkeley, CA: New Riders.

Elson, M., Breuer, J., Ivory J.D., and Quandt, T. (2014) More than stories with buttons: Narrative, Mechanics, and Context as determinants of player experience in digital games. *Journal of Communication* 64, 521–542.

Huizinga, J. (1955) *Homo Ludens*. Boston: The Beacon Press.

Hunicke, R., LeBlanc, M., and Zubek, R. (2004) MDA: A formal approach to game design and game research. Paper presented at the AAAI 19th National Congress on Artificial Intelligence, July 2004, San Jose, CA.

Ibister, K. (2009) Understand social play. In C. Bateman (ed.) *Beyond Game Design: Nine Steps Toward Creating Better Videogames* (pp. 49–59). Boston, MA: Charles River Media.

Jarvinen, A. (2002) Gran Stylissimo: The audiovisual elements and styles in computer and video games. In F. Mayra (ed.) *Proceedings of the Computer Games and Digital Cultures Conference* (pp. 113–128). 6–8 June 2002, Tampere, Finland.

Lang, A. (2000) The limited capacity model of mediated message processing. *Journal of Communication* 50 (1), 46–70.

Lazzaro, N. (2009) Understand emotions. In C. Bateman (ed.) *Beyond Game Design: Nine Steps Toward Creating Better Videogames* (pp. 3–48). Boston, MA: Charles River Media.

LeBlanc, M. (2004) Mechanics, Dynamics, Aesthetics: A formal approach to game design. Lecture at Northwestern University, April 2004. See algorithmancy.8kindsoffun.com/MDAnwu.ppt (accessed July 2016).

Lecky-Thompson, G.W. (2008) *Video Game Design Revealed*. Boston: Charles River Media.

Lindley, C.A. (2002) The gameplay gestalt, narrative, and interactive storytelling. In F. Mayra (ed.) *Proceedings of the Computer Games and Digital Cultures Conference* (pp. 203–215). 6–8 June 2002, Tampere, Finland.

Makkonen, T., and Hokkanen, T. J. (2013) ICT innovation and local economy: Mobile game as a tourist attraction. *Scandinavian Journal Of Hospitality & Tourism* 13 (3), 257–268.

Michael, D. and Chen, S. (2006) Serious games: Games that educate, train, and inform. Boston: Thomson.

Mitchell, B.L. (2012) *Game Design Essentials*. Indianapolis, IN: John Wiley & Sons.

Moore, M.E. (2011) *Basics of Game Design*. New York: CRC Press.

Norman, D.A. (1988) *The Design of Everyday Things*. New York: Basic Books.

Norman, D.A. (1999) Affordance, conventions and design. *Interactions* (May–June), 38–42.

Norman, D.A. (2004) *Emotional Design*. New York: Basic Books.

Oliver, M.B., and Raney, A.A. (2011) Entertainment as pleasurable and meaningful: Identifying hedonic and eudaimonic motivation for entertainment consumption. *Journal of Communication* 61, 984–1004.

Park, N., Lee, K.M., Jin, S.A. and Kang, S. (2010) Effects of pre-game stories on feelings of presence and evaluation of computer games. *International Journal of Human-Computer Studies* 68, 822–833.

Rusch, D.C. (2008) Emotional design of computer games and feature films. In A. Jahn-Sudmann and R. Stockman (eds) *Computer Games as a Sociocultural Phenomenon: Games Without Frontiers and War Without Tears* (pp. 22–31). New York: Palgrave Macmillan.

Salen, K. and Zimmerman, E. (2004) *Rules of Play: Game Design Fundamentals.* Cambridge, MA: The MIT Press.

Schell, J. (2008) *The Art of Game Design: A Book of Lenses.* New York: Morgan Kaufmann.

Schneider, E.F., Lang, A., Shin, M. and Bradley, S.D. (2004) Death with a story: How story impacts emotional, motivational, and physiological responses to first-person shooter video games. *Human Computer Research* 30 (3), 361–375.

Skalski, P., Tamborini, R., Shelton, A., Buncher, M. and Lindmark, P. (2011) Mapping the road to fun: Natural video game controllers, presence, and game enjoyment. *New Media and Society* 13 (2), 224–242.

Slak-Valek, N. and Williams, R.B. (2018) One place, two perspectives: Destination image for tourists and nationals. *Tourism Management Perspectives* 27 (2018), 152–161.

Thompson, J.J., Blair, M.R. and Henrey, A.J. (2014) Over the hill at 24: Persistent age-related cognitive-motor decline in reaction times in an ecologically valid video game task begins in early adulthood. *PLoS ONE* 9 (4), e94215.

Van der Post, L. (2010) *Jung and the Story of our Time.* New York: Random House.

Van den Berghe, J. (2012) The five domains of play. Presented at the Game Developers Conference, 5–8 May, San Francisco, CA. Available at: http://www.gdcvault.com/play/1015364/The-5-Domains-of-Play (accessed June 2016).

Williams, R.B. (2014) Game Design, Nine Elements and a Model. Paper presented to the Popular Culture Association / American Culture Association Conference, 14–16 March, Chicago, Illinois.

Williams, R.B. (2018) Conceptual models and mental models in operation: Frustration, performance and flow with two different video game controllers. *Entertainment Computing* 28 (2018), 2–10.

Williams, R.B. (forthcoming) Location, integration, interruption: Visual properties and recognition of video game advertising. *Journal of Promotion Management* 26 (2), 253–276.

Zimmerman, E. (2004) Narrative, interactivity, play, and games: Four naughty concepts in need of discipline. In N. Waldrip-Fruin and P. Harrigan (eds) *First Person: New Media as Story, Performance, and Game* (pp. 154–164). Cambridge, MA: MIT Press.

4 Gamification: Practices, Benefits and Challenges

Ye (Sandy) Shen and Marion Joppe

Introduction

According to the global research company Markets and Markets (2019), gamification in education will be worth US$1.8 billion by 2023 with a compound annual growth rate of 32% from 2018 to 2023. Zichermann and Cunningham (2011: xix) believe that '[b]aking gamification into your business can produce the ideal product.' With this great potential, gamification has been widely used in education, marketing and service design (Owen, 2013). In the past 10 years, several destinations and service providers have applied gamification in a tourism and hospitality context (Antonaci & Ott, 2014; Wang & Tsai, 2014; Xu *et al.*, 2016).

Destination Ontario (Canada), for example, launched a gamified advertisement, Where Am I, in English and French on 22 June 2016. Instead of offering the images of the province's landmark attractions, the advertisement showed a series of quick shots of its less familiar and surprising aspects, such as the magnificent hand-carved Swaminarayan Mandir of Toronto. At the end of the commercial, it asked, 'Where am I?' and directed viewers to WhereAmI.com to write their answers. The viewers were also provided with daily clues to figure out the riddle and a forum to guess the locations depicted in the spots. After 10 days, the advertisement revealed that the answer was Ontario and directed viewers to ontariotravel.net to discover the details of destinations depicted in the quick shots. More than 30,000 consumers participated in this riddle game, and the advertisement pulled in over 2.6 million views in 12 days (Martin, 2016a). It drew much more attention than previous campaigns. As noted by a number of authors (cf. Negruşa *et al.*, 2015), gamification offers a fun and exciting experience that results in better engagement. Although practitioners and researchers have discussed its promising future, market growth has been slower than expected. Therefore, there is a need to better understand the nature of gamification to improve its application. Additionally, it is crucial to examine previous gamification practices to have implications for future development. This chapter explains the benefits of applying gamification, discusses game mechanics as well as motivations to play, summarises gamification practices, and highlights some benefits and challenges of using gamification in a tourism and hospitality context.

Why Gamify a Non-Game Context

Gamification derives from the word 'game'. Understanding the concept of a game can provide insight into the nature of gamification. The definitions and characteristics of a game suggest why we need to gamify a non-game context. The *Oxford Dictionary* defines a game as 'an activity that one engages in for amusement or fun' and 'a complete episode or period of play, ending in a final result'. These definitions point to three characteristics. First, a game offers an enjoyable experience and triggers people's positive emotions, such as joy and interest. Second, a game needs people to spend some time to play and engage in the activity. Third, a game must have some kind of outcome (e.g. win or lose). In addition to these characteristics, researchers emphasise voluntary participation, certain rules and different parties (Avedon & Sutton-Smith, 1971). Abt (1968) defines a game as 'any contest (play) among adversaries (players) operating under constraints (rules) for an objective (winning, victory payoff)' (cited by Ellington *et al.*, 1982: 9). Different from previous studies, Juul's (2003) definition emphasises the positive impact of players' efforts on game outcomes. He describes a game as having six components, namely: rules; variable and quantifiable outcomes; values assigned to possible outcomes; players' efforts; players' attachment to outcomes; and negotiable consequences. Another critical component of a game is resolution (Seaborn & Fels, 2015), which means that players want to resolve challenges in games to gain a sense of achievement.

Based on these characteristics, gamifying a non-game context can have a number of benefits. Gamification offers an enjoyable experience and arouses positive emotions, which may lead to better satisfaction and improve loyalty (Xu *et al.*, 2014). Since the outcome of a game is uncertain, people make a greater effort to resolve challenges and to win, resulting in better engagement (Huotari & Hamari, 2012). A win and the accompanying sense of achievement further enhance their positive attitudes towards the gamified products or services. Specific and clear game rules provide guidance on completing the tasks in the gamified contexts and can motivate people to change behaviour (Barata *et al.*, 2013). For example, the rules for finding a hidden treasure and collecting coins in the gamified visitor guide Travel Plot Porto nudge tourists to visit more places in Porto, Portugal.

Increasing engagement is the primary purpose of gamifying a non-game context. The categories of games indicate different levels of challenges and engagement. There are six main game functions: recall and memorisation, judgement, consequence, strategy, exploration and simulation (Allen Interactions, 2014) (Figure 4.1). The games of recall and

Figure 4.1 Taxonomy alignment for gaming
Source: Allen Interactions (2014)

memorisation as well as judgement are fast-paced ones, where players quickly recall, match, recognise or identify correct answers. Examples include graphic identification, true/false and matching games. These games do not require much playing time, making participation easier for people. Since the interaction is relatively simple (e.g. answer questions, match cards), applying these two types may lead to less interest and fun. Conversely, consequence, strategy and exploration games require more effort from players who need to make decisions and explore diverse options to achieve different outcomes. This intensive involvement can create more enjoyable experiences. Finally, simulation games are the pinnacle of the taxonomy as they replicate activities in the real world. Since the contextual settings provide a sense of reality, players are more likely to become immersed in gaming environments. For instance, the augmented reality game Pokémon GO had attracted one billion downloads by September 2018, and its revenue was worth nearly US$200,000 per day as of April 2019, making it one of the most popular mobile games (Iqbal, 2020). Engagement levels are also affected by people's different interests and abilities to play (Robson *et al.*, 2016). Therefore, gamification providers should consider game types and their target groups' interests to engage users better.

Use of Game Mechanics by Gamification Providers

Product and service providers use game mechanics to gamify non-game contexts. From a narrow perspective, game mechanics are the features shared by games, while from a broad perspective, they are any elements that can be found in a game (Deterding *et al.*, 2011; Shen & Joppe, 2018). Game mechanics vary in different contexts with the mechanics of digital games being different from non-digital ones. For example, points, badges and levelling-up are common in role-playing videogames, but not in other types of games such as riddles (Robson *et al.*, 2016). Product and service gamification might involve only one or several of the listed game mechanics, but not necessarily all of them.

Game mechanics are a fundamental component of Hunicke *et al.*'s (2004) MDA (mechanics, dynamics, and aesthetics) model. They result in dynamic system behaviours and then lead to players' emotional responses (i.e. aesthetic experiences). Hunicke *et al.* (2004: 3) define game mechanics as 'the various actions, behaviours and control mechanisms afforded to the player within a game context'. Dynamics are a player's different actions or responses, such as competition, collaboration and sharing; aesthetics refers to a player's emotions including fun, excitement and surprise. According to the MDA model, appropriate game mechanics generate positive affection.

Game mechanics can be classified into three categories based on their different functions (Table 4.1). The first, game structure, refers to those that support the progress of a game, such as feedback, rules, progression and storyline. It organises games and allows them to proceed smoothly. The second, game tasks, include the different challenges and quests that people need to accomplish, for instance, goals, sharing tasks, quests and contests. These tasks challenge players and arouse their intrinsic interest. The third, game achievements, comprise the accomplishments or benefits of playing a game, such as points, badges, levels, leaderboard and rewards, heightening a player's engagement with the game and generating their extrinsic motivation to play.

Different game mechanics, especially those in the latter two categories, have been applied to gamification in destinations, hotels and restaurants (Shen & Joppe, 2018). Examples include consequence games (e.g. 4food Restaurants), strategy games (e.g. Fairy and the Safe) and exploration games (e.g. Strayboots). In the case of 4food Restaurants, customers can create their own type of burgers, name these creations and market them via social media. Both challenges and sharing game mechanics were used in this example. Fairy and the Safe is an augmented reality game employed by the Dutch theme park, Efteling, whereby visitors need to figure out a strategy to collect and save as many coins as possible by finishing different tasks while preventing the witch from stealing the coins. Here, challenges and rewards are both applied. Adopting the same game mechanics, Strayboots is essentially a mobile scavenger hunt that can focus on a specific attraction like a museum or a whole city, although it is used for team building rather than tourism.

Table 4.1 Game mechanics identified in previous studies

Examples	**Three categories of mechanics**		
	Game structure	**Game tasks**	**Game achievements**
Kankanhalli *et al.* (2012) (online platform)	Progression	Quests and challenges, viral loop	Points, virtual badges, leaderboard, levels and status
Xu *et al.* (2014) (tourism gamification)	Progress bar	Contests	Points, levels (gold, platinum, executive platinum), physical rewards badges, virtual currency, virtual goods
Wang and Tsai (2014) (website or application)	Feedback, rules	Challenges	Points, leaderboards, levels, badges
Bulencea and Egger (2015) (tourism gamification)	Feedback	Clear goals, sharing, collaboration	Points, rewards, leaderboards
Robson *et al.* (2016) (customer and employee engagement)	Infinite play, finite end, multiplayer orientation, team playing, online playing, real-world playing	Increasing task difficulty	Points, leaderboards, badges, new levels
Biel (2016) (physical activity)	Story/theme, progress, feedback	Challenge, clear goals	Points, leaderboards, achievements/ badges, levels, rewards

In terms of platforms, some gamification practices are based on digital ones (e.g. websites, smartphones and touchscreen tablets), while others are designed as activities. Most gamification practices of tourism destinations rely on digital platforms today. For example, location-based augmented reality mobile games are popular, as they can offer an immersive experience and result in better satisfaction (Koo *et al.*, 2018). Geocaching is also a new trend since it can provide tourists with enjoyable travel routes (Skinner *et al.*, 2018). Whether these game mechanics significantly trigger people's interest to play not only relies on game design but also depends on players' motivation.

Gamification Players: Motivation to Play

Understanding why tourists would like to engage in a gamified context can help designers understand player types, create valuable games and offer better experiences (Shen *et al.*, 2020; Xu *et al.*, 2016). Shen *et al.* (2020) categorised players of gamified trips into six types, including knowledge collectors, reward seekers, explorers, curiosity seekers, sensation seekers and flow experiencers based on Q-methodology. Most Generation Xers and Millennials are knowledge collectors, who are motivated by the epistemic value of a gamified trip while Generation Zers have relatively diverse interests. This finding has implications for how to design appealing gamified trips for different market segments (Shen *et al.*, 2020).

Motivation can be classified into extrinsic and intrinsic according to self-determination theory (Ryan & Deci, 2000). Extrinsic motivation occurs when people perform activities for known external rewards, including tangible (e.g. financial rewards) or psychological rewards (e.g. praise) (Brown, 2007). In gamification, points or badges in digital games or financial rewards and prizes in the real world, are used to generate extrinsic motivation. Gamified advertisements, like those produced by Mini Getaway Stockholm 2010, Pick N' Play and Coca-Cola's Chok Chok Chok campaign, used external rewards to attract consumers' attention. Specifically, the Mini Getaway Stockholm 2010 campaign designed a virtual reality game. People were challenged to chase and catch a virtual Mini automobile in Stockholm using an iPhone app. They then needed to run and protect it from being stolen by other players. The player who did so successfully over seven days won a real Mini. This campaign attracted 11,413 people from over 90 countries, and the virtual Mini was transported nearly 1500 kilometres. The average playing time per player was about 5 hours. This gamified advertisement was deemed a huge success with excellent consumer engagement.

However, some researchers criticise the use of external rewards to motivate consumers since they could effectively increase consumers' interest only for a short period but not necessarily over the long term (Werbach & Hunter, 2012; Xu *et al.*, 2017). Rewards are not applicable to all contexts. Companies (e.g. Starwood, BMW and Coca-Cola) and some destinations (e.g. museums) can provide the winners with discounts, merchandise or money. However, it is difficult for non-profit organisations that do not own any business, like destination management organisations (DMOs), to offer discounts or financial rewards.

Intrinsic motivation is argued to have a more significant effect than extrinsic motivation because it exists within individuals and links the relations between them and their activities more closely (Ryan & Deci, 2000). Xu *et al.* (2016) explored different types of intrinsic motivations to play a tourism-related game, including curiosity, exploring the destination, socialising, a fun and fantasy experience, challenge and achievement. For example, Ontario's domestic campaign, Where Am I, tried to satisfy the different levels of players' intrinsic motivation. This advertisement is a riddle game that triggered players' curiosity. People were able to explore the destination information on the website, read comments and have fun experiences. Players that correctly answered the riddle had a sense of achievement.

Gamification Practices, Benefits and Challenges

Gamification in tourism destinations

Several tourism destinations have undertaken gamification initiatives in recent years (Table 4.2), led by two main stakeholders: DMOs or tourism departments and IT companies. For the former, the aim is to attract more tourists by launching gamified promotions or offering gamified experiences. For example, the New Mexico Tourism Department wanted to entice visitors to vacation in-state with the gamified promotion New Mexico Catch the Kid. Tourists were motivated to visit different places to find clues about where Billy the Kid could be hidden. This promotion significantly nudged tourists to explore more places. Geocaching, as an outdoor recreational activity based on GPS, also effectively leads people to visit more locations. This type of game (e.g. The Pentati Pirate Trail Geocaching Treasure Hunt) satisfies people's motivation for seeking adventure and is especially attractive to Millennials and Generation Z (Skinner *et al.*, 2018). Another example is the Tourism Authority of Thailand's Smile Land Thailand. This videogame was designed based on the attractions in Thailand, aiming to trigger potential tourists' interest in visiting. The aforementioned Destination Ontario campaign, Where Am I?, had similar objectives.

Other initiatives, such as TravelPlot Porto, Strayboots, Tripventure and History Hero, were launched by IT companies. In these cases, destinations served as a backdrop for interactive gamified visitor guides. Players not only acquire travel information but also enjoy completing tasks in an engaging and fun way. For instance, tourists learn about the history of New York City by finishing challenges with the Strayboots game app. However, these IT companies do not offer free apps since their income derives from people's purchases. Although the games themselves were exciting and engaging, some people were reluctant to purchase them, especially those that they can only be played at the destination. As a result, these companies did not attract sufficient players and with the exception of Strayboots, are not active or ceased to exist.

Gamification in tourism destinations is becoming popular as it benefits both DMOs (or IT companies) and tourists. For DMOs, it can provide instant information, build a positive destination image, offer an enjoyable experience, increase satisfaction and nudge visitors to explore more places. This can lead to increased visitation and economic returns on investment. For tourists, the information presented in a gamified way enables them to learn more about the destination. For example, the gamified tour History Hero primarily targets children and motivates them to gain knowledge by visiting museums. Additionally, a gamified tour has been added with more hedonic value that makes tourists feel that their trip is worthwhile.

Table 4.2 Gamification in destinations

Destinations	Gamification examples	Resources	Stakeholders	
			Gamification providers	Gamification players
New Mexico, USA	*New Mexico Catch the Kid (2011)*: Visitors were invited to seek different locations where Billy the Kid could be hidden. The winner, who followed the clues to track down the bandit Billy the Kid and bring him to justice, received US$10,000 from the New Mexico Tourism Department.	Miller (2011)	New Mexico Tourism Department	Visitors in New Mexico
Thailand	*Smile Land Thailand (2011)*: Players could learn about Thailand at a glance in an amusing way. By playing the videogame, players discover hidden treasures and particular tourist spots in Thailand.	CNN Travel (2011); Xu *et al.* (2016)	Tourism Authority of Thailand	Visitors interested in Thailand
Regensburg, Germany	*REXplorer (2007)*: Visitors could visit different places while playing the location-based spell-casting game.	Walz and Ballagas (2007)	Regensburg Experience Museum	Visitors in Regensburg
Denton, USA	*I Spy Denton (2012)*: Visitors were invited to complete tasks and challenges and interact with the destinations.	Negruşa *et al.* (2015); TTIA (2012)	TTIA's Travel & Tourism College	Visitors in Denton
Dutch Theme Park Efteling, Netherlands	*Fairy and the Safe (2011)*: Visitors were invited to play an augmented reality game. Players needed to collect and save as many coins as possible by finishing different tasks and preventing the witch from stealing the coins.	Digital Tourism Think Tank (2014)	Dutch Theme Park Efteling	Children visiting the Dutch Theme Park Efteling
Corfu, Greece	*The Pentati Pirate Trail Geocaching Treasure Hunt (2015)*: In this geocaching game, players can use a GPS to seek treasures in containers and explore Pentati Village.	Skinner *et al.* (2018)	Pentati Village	Visitors to Pentati Village

Porto, Portugal	*TravelPlot Porto (2012)*: Visitors can win rewards if they accomplish the tasks (e.g. identify various hidden treasures). They can also share their stories related to the destination with friends on Facebook or Twitter.	Digital Tourism Think Tank (2014); Negruşa *et al.* (2015)	TravelPlot Porto	Visitors interested in Porto
New York City, USA	*Strayboots (2012)*: Visitors follow the electronic map to finish various tasks and discover hidden spots and different places.	Digital Tourism Think Tank (2014); Etherington (2012)	Strayboots	Visitors to New York City
Stockholm, Berlin, Hamburg, Cologne, London, Munich, Rome, Paris and Barcelona	*Tripventure (2012)*: This platform offers several location-based augmented reality games. Visitors are invited to meet virtual characters, solve riddles, find objects and visit landmarks. These games were designed as a storyline.	Tripventure (2012)	Tripventure	Visitors to Stockholm, Berlin, Hamburg, Cologne, London, Munich, Rome, Paris and Barcelona
Metropolitan Museum of Art, St Peter's Basilica, Versailles Castle and National Mall	*History Hero (2013)*: Visitors are invited to play an interactive videogame museum guide. They gain points by finishing the missions (e.g. answer questions and take photos).	Wang and Tsai (2014)	History Hero	Children visiting the museums and historic sites

Even though gamification has been applied to tourism destinations and can bring some benefits, it has experienced relatively slow growth. Previous initiatives were launched mainly in 2011 and 2012, when the concept of gamification was first introduced. However, there have been very few representative practices in recent years. The application of gamification in tourism destinations is stunted by some challenges.

First, the main reason for gamification is to make tedious or boring contexts more interesting, but tourism destinations are already offering fun and engaging experiences, which reduces the need for gamification. Therefore, providers should better understand tourists' needs to gamify tourism destinations effectively. For example, Strayboots has morphed into a service company offering hunt experiences for team building, employee engagement and orientation. Players are invited to form groups and complete challenges at destinations. This kind of hunt game is especially appropriate for destinations with which players are familiar, as familiarity reduces visit interest. However, gamification makes a trip more exciting and leads people to be more interested in learning about the place.

Second, tourists have limited time when visiting a destination while playing games also requires time, so some people are not willing to take a gamified tour or engage with a gamified visitor guide. They would like to allocate more time to experience the destination itself. Thus, gamifying a place where people live or with which they are familiar would be a better alternative. Since they have plenty of time at that place, making it more interesting is meaningful. A good example is the popularity of Pokémon GO. Players are keen to exercise or hang around familiar places in the hopes of catching Pokémon. Although Pokémon GO has also been used by DMOs to promote tourism, the number of tourists motivated to go to a destination to collect an uncommon Pokémon remains limited (Diaz, 2020).

Third, some gamification practices rely on digital platforms. However, when people visit a destination, they prefer to look at the scenery or historic sites instead of focusing on a screen. It is challenging to integrate tourist attractions into the virtual world of a digital game, especially when Wi-Fi availability is constrained or battery time is limited. Although augmented reality games are innovative and offer playful experiences, some tourists are reluctant to engage with them because they believe it impedes the authentic travel experience. Overall, gamifying tourism destinations through digital platforms needs careful thought.

Gamification in tourism and hospitality marketing

Gamification has also been applied to tourism and hospitality marketing, where it aims to support branding and marketing goals (Madlberger *et al.*, 2018). Destination Ontario launched several gamified advertisements, such as Where Am I, Fantastic Ontario Family Trip and Ontario

Colourful Spring Tour (Table 4.3). Instead of showing the iconic scenic spots and attractions of the province, the advertisements invited viewers to participate in games and then provided travel information. This approach successfully attracted audiences and increased advertising engagement. Another good example is McDonald's Pick N' Play. People in Stockholm actively participated in the game to win free treats. When this game was made available on YouTube, it received more than 0.6 million clicks, making it quite a successful promotional campaign. Marketers

Table 4.3 Gamification in advertising

Gamified ad	Description	Stakeholders	
		Gamification providers	**Gamification players**
Where Am I (2016)	People were invited to play a riddle game and guess which destination Where Am I is describing. They could provide their answers at www.whereami.com and browse travel information.	Destination Ontario	Ontario residents
Ontario Colourful Spring Tour (2016)	This advertisement offered a story narrative. People could explore four Ontario spring experiences (i.e. Ottawa Tulip Festival, Toronto Sakura Blossoms, Niagara Falls Butterfly Museum and Blue Mountain Village) and share their favourite destinations on social media to win printed game gifts.	Destination Ontario	Chinese interested in visiting Ontario
Fantastic Ontario Family Trip (2016)	People were invited to answer online questions to create customized five-day family vacation itineraries and predict the parenting style of users.	Destination Ontario	Chinese families interested in visiting Ontario
Austria Snapshot Adventure (2015)	In this advergame, people could design their avatars, travel to three destinations in mini games (such as Tyrol, Salzburg and Vienna), take pictures and share with friends.	Austria Tourism	Potential tourists
Pick N' Play (2011)	People were invited to play an interactive outdoor campaign. Games were played on a large interactive billboard by using their mobile phones. The winners received their favourite McDonald's treat.	McDonald's	People in Stockholm
Halloween Let's Play (2016)	People were offered a mixed-reality game experience. The Halloween game was designed to promote Burger King's products. The winners received discount coupons and other rewards.	Burger King	Digital game players

Source: Shen and Joppe (2018)

should consider videogames as a sufficient way to motivate potential tourists since experiencing virtual attractions significantly influences affection, impacts cognition and triggers people's visit interest (Dubois & Gibbs, 2018; Madlberger *et al.*, 2018). In Madlberger *et al.*'s (2018) study, participants expressed feeling fun, curiosity, joy and happiness after playing Austria Snapshot Adventure.

Gamified advertisements can also help service providers to understand better which products or services are more or less attractive to audiences. For example, Destination Ontario learned from its Ontario Colourful Spring Tour that the Ottawa Tulip Festival and Toronto's Sakura Blossoms held the greatest appeal for users (Shen & Joppe, 2018). This is a better way than conducting surveys to know which destinations or products are more popular.

Viral marketing can be another benefit, especially when audiences participate in the games and share information, such as in the marketing campaigns of Fantastic Ontario Family Trip and Austria Snapshot Adventure. Brand information will spread to social media where it can be reposted. It will bring even more attention to that brand and raise people's brand awareness.

The application of gamification in marketing is trendy with most of the listed gamification practices initiated in 2016. However, this application is also associated with some challenges. From the perspective of gamification providers, whether players actually pay attention to a brand rather than just playing the game is questionable. Researchers have investigated how to better integrate brands into advergames (Vashisht & Royne, 2016) and found that placing a brand in a low-speed advergame is more effective than doing so in a high-speed one. Brand placement also plays a crucial role in generating brand recall: the more visible a brand is, the higher the resultant brand recall. Another challenge is that people's engagement with the gamified ad is limited since each one is relatively short. The Ontario Colourful Spring Tour takes two minutes to complete, and McDonald's Pick N' Play is even shorter at only 30 seconds. This short play time may lead to lower involvement. Extending player participation time contributes to a higher level of engagement.

Gamification to combat overtourism

Gamification has also been deemed effective in leading tourists to behave more sustainably (Yoo *et al.*, 2017). It nudges people to choose public transportation, stay in certified sustainable accommodation, save energy and purchase local food or products (Negruşa *et al.*, 2015). Some destinations have adopted gamification to tackle overtourism (Table 4.4). For instance, Visit London launched Play London with Mr Bean, where players can receive discounts and vouchers for attractions, restaurants and shops. This game aims to disperse tourists by encouraging them to visit

Table 4.4 Gamification to address overtourism

Destinations	Gamification examples	Resources	Stakeholders	
			Gamification providers	**Gamification players**
London, UK	Play London with Mr Bean (2018): a match-three puzzle videogame. Players can receive points by completing each level and then redeem discounts and vouchers around the city.	Peltier (2018), WTTC (2018)	Visit London	Visitors interested in London
Los Angeles, USA	A Catch List of Pokémon (2016): People were invited to visit different places in Los Angeles to catch uncommon Pokémons and enjoy Pokémon-related discounts at attractions and restaurants.	Martin (2016b)	Los Angeles Tourism & Convention Board	Visitors in Los Angeles
Vail Resorts, USA	EpicMix: Skiers and riders can complete different levels of on-mountain challenges, receive badges and find out which chairlift lines are shortest to avoid congestion.	Ladyga (2010)	Vail Resorts and General Motors Truck Company (GMC)	Visitors in Vail Resorts

some lesser-known places across the city, such as Greenwich. Los Angeles published a list of spots where players can catch less common Pokémons to attract players and receive Pokémon-related discounts at attractions and restaurants (Martin, 2016b). This initiative is useful to direct visitors to different locations, especially less popular or less known ones (Diaz, 2020). The fun interaction of gamification enriches tourists' experiences and adds more value to those destinations that receive fewer visitors.

Using gamified apps to offer better information for tourists is also a promising way to control crowds. For example, skiers can use EpicMix to check lift line wait times in Vail Resort to avoid congestion. This gamified app not only disperses tourist flows but also enhances their experiences. Efforts made by organisations to do just that, like the My Disney Experience mobile app, which provides attraction wait times and allows visitors to design their visit to avoid crowds, could be made even more useful.

Gamification is a promising solution for overtourism by spreading visitation across destinations and controlling tourist flows. Through gamified apps, tourists can visit hidden travel destinations and share experiences others have never had, something that is particularly appealing to Generations Y and Z who are more interested in these types of experiences (Joppe, 2019) and keen to brag about their discoveries (Kerr *et al.*, 2012; Nusair *et al.*, 2013). Additionally, they can monitor wait times and strategically plan their routes. Although applying gamification seems to address overtourism effectively, it is not without its challenges. Since these apps are on a digital platform and service providers need to publish real-time information, they do not reach the visitors who do not use apps at all. But gamification can be useful in controlling the crowds of Generations Y and Z, who are the heaviest users of mobile apps.

Conclusions

This chapter explains why and in which contexts it is essential to apply gamification. Three categories of game mechanics (i.e. game structure, game tasks and game achievements) are discussed from the perspective of how they trigger players' intrinsic or extrinsic motivations to play. Intrinsic motivation is believed to play a more important role since it reflects people's inner desires.

The analysis of gamification practices in the tourism and hospitality context identified organizations responsible for tourism marketing (e.g. tourism departments and DMOs) and IT companies as the two main gamification providers. The former offer free gamified experiences to attract more visitors while the latter charge players since their objective is to derive income from providing gamified visitor guides. These two types of providers are both facing some challenges. Gamification is meant to add a level of interest and engagement to what may be perceived as a relatively boring or mundane context. Since visiting a destination is generally already an enjoyable experience, adding value through gamification can be a big challenge. The use of a digital platform can also interfere with the actual travel experience at the destination. The good practices discussed in this chapter should provide some ideas of how these challenges can be overcome, enhance the future application of gamification, and how it might assist in dispersing tourist flows and controlling crowds.

References

Abt, C.C. (1968) Games for learning. In S.S. Boocock and E.O. Schild (eds) *Simulation Games in Learning* (pp. 65–84). Beverly Hills, CA: Sage Publications.

Allen Interactions (2014) Taxonomy alignment for gaming. *Twitter.* Available at: https://twitter.com/customelearning/status/481489023509094402/photo/1 (accessed 2 July 2019).

Antonaci, A. and Ott, M. (2014) Virtual museums and learning innovation. In I. Roceanu (ed.) *Let's Build the Future through Learning Innovation* (pp. 28–33). Bucharest: The Carol I National Defence University.

Avedon, E.M. and Sutton-Smith, B. (1971) *The Study of Games.* New York: John Wiley & Sons.

Barata, G., Gama, S., Jorge, J. and Goncalves, D. (2013) Engaging engineering students with gamification. Paper presented at the 5th International Conference on Games and Virtual Worlds for Serious Applications (VS Games), 11–13 September, Bournemouth, UK.

Biel, A.M. (2016) Pokémon GO: A socio-technical exploratory study. Master's thesis, Arizona State University.

Brown, L.V. (2007) *Psychology of Motivation*. New York: Nova Science Publishers.

Bulencea, P. and Egger, R. (2015) *Gamification in Tourism: Designing Memorable Experiences*. Norderstedt: Books on Demand.

CNN Travel (2011) Smile Land: TAT launches social media game apps. *CNN Travel*. Available at: http://travel.cnn.com/bangkok/play/smile-land-thai-tourism-launches-social-media-game-apps-774959/ (accessed 1 April 2020).

Deterding, S., Dixon, D., Khaled, R. and Nacke, L. (2011) From game design elements to gamefulness: Defining 'gamification'. Paper presented at the 15th International Academic MindTrek Conference: Envisioning Future Media Environments, 28–30 September, Tampere, Finland.

Diaz, J. (2020) Pokémon GO helped generate $249M in tourism revenue in 2019. *Android Headlines*. Available at: https://www.androidheadlines.com/2020/01/pok%C3%A9mon-go-helped-generate-249m-in-tourism-revenue-in-2019.html (accessed 1 April 2020).

Digital Tourism Think Tank (2014) Gaming and gamification in tourism: 10 ways to make tourism more playful. Available at: https://thinkdigital.travel/wp-content/uploads/2014/05/Gamification-in-Tourism-Best-Practice.pdf (accessed 2 July 2019).

Dubois, L.E. and Gibbs, C. (2018) Video game-induced tourism: A new frontier for destination marketers. *Tourism Review* 73 (2), 186–198.

Ellington, H., Addinall, E. and Percival, F. (1982) *A Handbook of Game Design*. London: Kogan Page.

Etherington, D. (2012) Location-based tour guide App Stray Boots raises $2M as it looks to expand mobile reach. Available at: https://techcrunch.com/2012/11/27/location-based-tour-guide-app-stray-boots-raises-2m-as-it-looks-to-expand-mobile-reach/ (accessed 2 July 2019).

Hunicke, R., LeBlanc, M. and Zubek, R. (2004) MDA: A formal approach to game design and game research. Paper presented at the 19th National Conference of Artificial Intelligence, 25–29 July, San Jose, California.

Huotari, K. and Hamari, J. (2012) Defining gamification: A service marketing perspective. Paper presented at the 16th International Academic MindTrek Conference, 3–5 October, Tampere, Finland.

Iqbal, M. (2020) Pokémon GO revenue and usage statistics (2019) *Business of Apps*. Available at: https://www.businessofapps.com/data/pokemon-go-statistics/ (accessed 3 March 2020).

Joppe, M. (2019) The roles of policy, planning and governance in preventing and managing overtourism. In R. Dodds and R.W. Butler (eds) *Overtourism: Issues, Realities and Solutions* (pp. 250–261). Berlin, Germany: De Gruyter.

Juul, J. (2003) The game, the player, the world: Looking for a heart of gameness. Paper presented at the 1st International Digital Games Research Conference, 4–6 November, Utrecht, the Netherlands.

Kankanhalli, A., Taher, M., Cavusoglu, H. and Kim, S.H. (2012) Gamification: A new paradigm for online user engagement. Paper presented at the 33rd International Conference on Information Systems, 16–19 December, Orlando, USA.

Kerr, G. M., Lewis, C. and Burgess, L. (2012) Bragging rights and destination marketing: A tourism bragging rights model. *Journal of Hospitality and Tourism Management* 19 (1), 7–14.

Koo, C., Choi, K., Ham, J. and Chung, N. (2018) Empirical study about the PokémonGo game and destination engagement. In B. Stangl and J. Pesonen (eds) *Information and Communication Technologies in Tourism 2018* (pp. 16–28). Cham, Switzerland: Springer International Publishing.

Ladyga, K. (2010) Vail resorts announces EpicMix™ a new online and mobile application, allowing skiers and riders a seamless and effortless way to digitally capture and share their mountain experience. Available at: http://news.vailresorts.com/corporate/

vailresorts/vail-resorts-announces-epicmix-new-online-and-mobile-application-allowing-skiers-and-riders-seamless-and-effortless-way-to-digitally-capture-and-share-their-mountain-experience.htm (accessed 2 July 2019).

Madlberger, M., Stipetic, M. and Dlacic, J. (2018) Forming affect, behavior, and cognition with gamification: The case of an Austrian tourism advergame. Paper presented at Multikonferenz Wirtschaftsinformatik, 6–9 March, Lüneburg, Germany.

Markets and Markets (2019) Gamification in education market. Available at: https://www.marketsandmarkets.com/Market-Reports/gamification-education-market-10910763.html (accessed 3 March 2020).

Martin, R. (2016a) Ontario obscures identity for teaser tourism campaign. *Marketing*. Available at: http://www.marketingmag.ca/advertising/ontario-obscures-its-identity-for-teaser-tourism-campaign-178903 (accessed 3 July 2019).

Martin, L. (2016b) Top 5 unique ways destinations are using Pokémon GO. *Destination Think*. Available at: https://destinationthink.com/destinations-using-pokemon-go/ (accessed 3 July 2019).

Miller, L. (2011) New Mexico Tourism launches Billy the Kid - Themed scavenger hunt featuring $10,000 reward. *Gadling*. Available at: http://gadling.com/2011/07/29/new-mexico-launches-billy-the-kid-themed-scavenger-hunt-for-10/ (accessed 3 July 2019).

Negruşa, A.L., Toader, V., Soficş, A., Tutunea, M.F. and Rus, V.R. (2015) Exploring gamification techniques and applications for sustainable tourism. *Sustainability* 7 (8), 11160–11189.

Nusair, K., Bilgihan, A. and Okumus, F. (2013) The role of online social network travel websites in creating social interactions for Generation Y travelers. *International Journal of Tourism Research* 15 (5), 458–472.

Owen, P. (2013) *How Gamification Can Help Your Business Engage in Sustainability*. Shipley, UK: Greenleaf Publishing.

Oxford Dictionary. (n.d.). Game. Available at: https://www.lexico.com/definition/game (accessed 19 December 2020).

Peltier, D. (2018) London uses mobile gaming app to help tackle overtourism. *Skift*. Available at: https://skift.com/2018/06/27/london-uses-mobile-gaming-app-to-help-tackle-overtourism/ (accessed 3 July 2019).

Robson, K., Plangger, K., Kietzmann, J.H., McCarthy, I. and Pitt, L. (2016) Game on: Engaging customers and employees through gamification. *Business Horizons* 59 (1), 29–36.

Ryan, R.M. and Deci, E.L. (2000) Self-determination theory and the facilitation of intrinsic motivation, social development, and well-being. *American Psychologist* 55 (1), 68–78.

Seaborn, K. and Fels, D.I. (2015) Gamification in theory and action: A survey. *International Journal of Human-Computer Studies* 74, 14–31.

Shen, Y. and Joppe, M. (2018) Gamification in tourism advertising: Game mechanics and practices. *e-Review of Tourism Research*. Available at: https://ertr.tamu.edu/files/2018/01/ENTER2018_Submission_84-ok.pdf (accessed 3 July 2019).

Shen, Y., Choi, H.C., Joppe, M. and Yi, S. (2020) What motivates visitors to participate in a gamified trip? A player typology using Q methodology. *Tourism Management* 78. DOI: https://doi.org/10.1016/j.tourman.2019.104074.

Skinner, H., Sarpong, D. and White, G.R. (2018) Meeting the needs of the Millennials and Generation Z: Gamification in tourism through geocaching. *Journal of Tourism Futures* 4 (1), 93–104.

Tripventure (2012) Tripventure – location based games and guides. Available at: https://www.youtube.com/watch?v=fvZIMH0M3bA&feature=youtu.be (accessed 3 July 2019).

TTIA (Texas Travel Industry Association) (2012) TTIA's Travel & Tourism College I Spy Denton. Available at: https://www.ttia.org/events/EventDetails.aspx?id=190558 (accessed 3 July 2019).

Vashisht, D. and Royne, M.B. (2016) Advergame speed influence and brand recall: The moderating effects of brand placement strength and gamers' persuasion knowledge. *Computers in Human Behavior* 63, 162–169.

Walz, S.P. and Ballagas, R. (2007) Pervasive persuasive: A rhetorical design approach to a location-based spell-casting game for tourists. Available at: http://www.digra.org/wp-content/uploads/digital-library/07311.17007.pdf (accessed 3 July 2019).

Wang, Y.C. and Tsai, P.H. (2014) A study on the applications of the cultural history promotion via gamification based. Digital curation: Taking 'History Hero' as an example. Available at: http://140.109.14.50/document/2014%20Poster_Third%20Place_Poster%20Image.pdf (accessed 3 July 2019).

Werbach, K. and Hunter, D. (2012) *For the Win: How Game Thinking Can Revolutionize Your Business*. Philadelphia, PA: Wharton Digital Press.

WTTC (World Travel & Tourism Council) (2018) How can new technologies help deal with overcrowding? Available at: https://medium.com/@WTTC/how-can-new-technologies-help-deal-with-overcrowding-dda554fb164b (accessed 3 July 2019).

Xu, F., Weber, J. and Buhalis, D. (2014) Gamification in tourism. In Z. Xiang and I. Tussyadiah (eds) *Information and Communication Technologies in Tourism* (pp. 525–537). Cham, Switzerland: Springer International Publishing.

Xu, F., Tian, F., Buhalis, D., Weber, J. and Zhang, H. (2016) Tourists as mobile gamers: Gamification for tourism marketing. *Journal of Travel & Tourism Marketing* 33 (8), 1124–1142.

Xu, F., Buhalis, D. and Weber, J. (2017) Serious games and the gamification of tourism. *Tourism Management* 60, 244–256.

Yoo, C., Kwon, S., Na, H. and Chang, B. (2017) Factors affecting the adoption of gamified smart tourism applications: An integrative approach. *Sustainability* 9 (12), 2162. doi: https://doi.org/10.3390/su9122162.

Zichermann, G. and Cunningham, C. (2011) *Gamification by Design: Implementing Game Mechanics in Web and Mobile Apps*. Sebastopol, CA: O'Reilly Media.

Part 2

Gamification Application and Case Studies

5 Gamification Applications in Hospitality and Airline Industries: A Unified Gamification Model

Zuhal Cilingir Uk and Yaşar Gultekin

A game is a series of interesting choices.
Sid Meier

In this chapter, already employed applications of gamification in tourism are presented and discussed. This chapter also aims to present a brief history of gamification along with a theoretical framework from the literature review. Gamification examples from the hospitality and airline industries are given and analysed in terms of various factors (game elements used, target audience, benefits, time period and involved core drives). By adopting the strengths of these applications and assessing the probable opportunities and best practice, a unified gamification model is offered.

A Brief History of Gamification

Gamification is the commonly used term to describe using game design elements and gameplay mechanics in non-gaming systems. Firms use gamification in order to enhance their customer loyalty programmes as part of their customer relations management (CRM). Numerous enterprises and institutions such as sport clubs and educational institutions (e.g. universities, institutes, colleges and so on) have developed their own gamification applications to strengthen their marketing campaigns. Hospitality and airline industries also use gamification to ensure brand loyalty and develop communication. They achieve customer engagement by creating successful guest experiences via gamification applications. Gamification is not a new concept to the hospitality and airline industries, but using the emerging digital tools and social media requires new, innovative and holistic approaches. The relation between new media and gamification applications is focused on promoting and managing the app in social media. Conducting the right mixture of gaming experience in a social network raises awareness and participation. But that does not happen in one day.

Human–computer interaction (HCI) has evolved dramatically and drastically from its early days, to the point that humankind and human mind evolved to get the ability to be a source of a so-called life to create AI besides basic algorithms. HCI is a field of study that has become more important than ever at any point in its history. Huizinga (2014) described our species as *Homo ludens* (playful human) in the 1930s. At the beginning of the 1980s, Malone published his work on the features that make games more engaging when used to make learning interesting and enjoyable (Malone, 1981). When the popularity of gamification is examined, it can be seen that it started to gain popularity at the beginning of 2011 and reached its peak point at the end of 2012, and has sustained the same level of popularity until today, albeit with minor fluctuations.

The notions of 'ludic design', 'ludic engagement' and 'ludic activities' were introduced by Gaver *et al.* (2004). Gamification is the process of game-thinking and game mechanics to engage users and solve problems (Zichermann & Cunningham, 2011) and the use of game design elements in non-game contexts (Deterding *et al.*, 2011). Scholars regard games as something different from 'play' (Caillois, 2001: 29). The term *paidia* (play) refers to a non-obligatory, expressive, uncertain and unproductive form of play; whereas, the term *ludus* (game) is the form of paidia with predetermined rules and goals. It is certain that paidia is the main constituent and sine qua non to creating ludus. Likewise, while determining the relationship between 'games' and 'play', Salen *et al.*, (2004) proposed that games are a subset of play as games can be described as play that possess a formalised set of rules and players compete to win. They have also proposed that play is a component of games, as the playing of the game represents one aspect of games.

Gameful design is differentiated from playful design on the ludus–paidia axis and resembles serious games on the parts–whole axis. For instance, according to these classifications, chess is defined as a serious game but a teddy bear falls under the category of toys that are part of a child's play. To understand better, the use of toys with no purpose other than playing or in playing in part is called playful design (affording paidic qualities) or toyification.

The Concept and Benefits of Gamification

Game practices are differentiated from gameful design. In contrast to 'serious games', gamified applications only use some elements of games (Groh, 2012: 40). Elements of games may be found outside the games environment, so the term design is used instead of game-based practices. While a standard game may or should use all levels of game design elements, a gamification application merely uses several design elements.

Gameful design or gamification shares a common conceptual space with playful design by using game elements partially rather than as a whole. Gamification is also similar to serious games as they are both determined as 'gaming' rather than 'playing'. It is also noteworthy that an application may be called either 'a game' or 'a gamified application' according to the users' or designers' intentions. For example, the popular application, Swarm, may be both used and played by users, as someone may use the application to socialise and get in touch with people in the first instance while someone else may use the application to win badges and to be at the top of the leaderboard.

Werbach and Hunter (2012) proposed a classification covering the target audience and the scope of the benefit. They determined whether the target audience is internal or external and the desired behaviour change is either for individuals or communities. Internal gamification, also called enterprise gamification, aims to improve productivity, enhance the commitment of employees to the company and achieve positive results for the corporation through this process. In internal gamification, it is crucial that the gamification application is compatible with an existing reward system as well as the management philosophy and needs.

External gamification targets customers and/or potential customers as part of the marketing efforts. This type of gamification can be seen as a way to build a stronger relationship between customers and the company in order to increase engagement, a higher level of loyalty and as a result higher revenue and profit. A gamification application also may aim to cause a desired behavioural change both in individuals and enterprise programmes. This kind of gamification is often applied by non-profit organisations and governments. They are generally used for health and educational purposes that affect entire communities.

Gamification Design

Werbach and Hunter (2015) also proposed a hierarchical classification of game elements that consisted of three categories from the top to down, that is, dynamics, mechanics and components (https://www.researchgate.net/figure/The-hierarchy-of-game-elements-Werbach-Hunter-2012_fig1_283469412). Dynamics are the category that consists of elements that do not directly enter into game and are conceptual elements, namely: constraints, emotions, narrative, progression and relationships. Constraints describe the limitations of the design that can be excluded. Emotions are the drives for user engagement of into the game. Each user is unique in terms of drives and that should be considered when determining the emotions. Narrative can be described as the storyline of the game and it is used to capture the attention and draws the borders of the users' interaction with the game. Progression refers to the development of the users' game experience via themselves or the avatar they have created while playing. Naturally, a player's understanding and skills in the game develop or the character created has better statistics through time. Relationships are the social interactions that users engage in during gameplay. These relationships can be with other people or in-game non-player characters (NPCs).

Mechanics are the main processes that make people play the game and get in touch with it. There are 10 of these mechanic elements: challenges, chance, competition, cooperation, feedback, resource acquisition, rewards, transactions, turns and win states. Challenges are obstacles that players are expected to overcome within the scenario of the game. The difficulty level of challenges is also important. If the level is not suitable to the progress made by the player (whether it is low or high), players may lose interest in the game. Chance refers to the randomness of elements in a game.

If the game or gamification application is predictable in terms of what will happen on the next step or if it repeats itself, it will be hard to keep users engaged in the game. Competition is the situation where some win and some lose. In game context, competition should be fair although it should carry some element of randomness. There should be a balance between chance and fairness. Cooperation is when users work together to achieve a goal in order to win. Gameplay has to allow for cooperation between users. Feedback is the sum of the data about the users' playing habits and patterns.

Resource acquisition is somehow unique to the gameplay, as it does not always happen in real-life situations. As the players progress in the game, they acquire items that are useful within the gameplay. For many, the acquisition is more meaningful than purchasing it, as they feel themselves rewarded with the progress they have made. Affiliated with that, rewards are benefits that players receive as a result of success within the game. Rewards either can be somewhat 'tangible' resources and useful in game or titles used in gameplay that differentiates players from others. Transactions are actions of trade taken between players. These transactions can be performed directly or with the use of an intermediary. Turns, although not be generally applicable to all kinds of games except turn-based ones, are the series of participation for each player. Win states represent the conditions that define the winner of a game as well as the ones who lose.

Nicholson (2015) offers the use of six elements (play, exposition, choice, information, engagement and reflection) inspired by game design in order to have a meaningful gamification application. Of these elements, play means the ease of the freedom to explore, exposition means the integration of the created stories with the real world and choice denotes the freedom of choice that the players should have. Information is the key

Table 5.1 Game elements

Visual elements	**Ranking elements**	**Socialisation elements**
Collections Avatars Badges	Leaderboards Levels Points	Social graphs Teams Gifting
Challenge elements	**Conferment elements**	**Scenario elements**
Boss fights Combat	Content unlocking Rewards Virtual goods	Story/theme Clear goals Progress

concept for learning, engagement means the ability of the application to let players interact with each other and reflection enables deepening engagement and learning through reflecting players' in-game experience to real life or vice versa.

Components are the most populated subsets of game elements. Although the most used game elements are points, leaderboards and badges (Hamari & Eranti, 2011), some other components can be seen in Table 5.1 (Hamari *et al.*, 2014; Werbach & Hunter, 2015).

Game elements can be classified into six categories. These are visual, ranking, socialisation, challenge, conferment and scenario elements. Collections (sets of items or badges collected in game), avatars (an icon or figure representing a particular person in a digital environment, i.e. gameplay) and badges (a special or distinctive mark, token or device worn as a sign of allegiance, membership, authority, achievement, etc.) are visual elements. Ranking elements are the tools that enable players to determine their status in the game. Ranking elements consist of leaderboards that display the names and current scores of the leading competitors in terms of game units such as points, badges, achievements, etc., levels that are pre-defined steps of in-game progression and lastly points that numerically represent what a player has acquired in terms of game units. Socialisation elements such as teams, gifting and social graphs enable the application to fulfil the socialisation needs of users. Social graphs demonstrate the players' social network within the game as they unite in a team and have the opportunity to share in-game items or resources with team mates through gifting.

Conferment elements are virtual goods (game assets), rewards determined as anything given in exchange for a requested behaviour or completed tasks and content unlocking featuring specialties or items available for the player when they have achieved a specific objective in the game. Combat and boss fights (hard challenges players have to face in the final game chapters) are challenge elements that empower players emotionally through the enjoyment of achieving a goal and the illusion of having overcome obstacles. Scenario elements such as story/theme, clear goals and progression made during the game develop the playability and ensure the continuity of the game.

Gamified applications may not use all levels of these elements, but in the context of the needs of the users and designers, some elements should be used to make an application gamified. Martens and Mueller (2016) simplified levels in game design and suggested three different levels: game elements, game mechanics and game thinking. In the model they proposed, game elements have two subgroups: interaction elements and rating elements. Interaction elements are defined as everything a player interacts with in game, such as NPCs, the virtual world, and so on. Rating elements are defined as every aspect of the game that gives feedback to the player. Game mechanics are differentiated into three subcategories: rules, roles and stories. Finally, game thinking can be understood as 'the use of games and game-like approaches to solve problems and create better experiences' (Sicart, 2008).

Drivers of Gamification

As well as designing the 'game', analysing the drivers of the gamification process is also crucial to obtain successful results. Some analytical techniques are offered to evaluate different aspects of the process. In Figure 5.1, Yu-kai Chou (2015) proposed a technique called Octalysis in which eight core drives of gamification are determined, graphed in an octagon chart.

The eight core drives that he proposed include (Chou, 2015):

(1) Epic meaning and calling: when a player has the thought that they are doing something greater than themselves and believes that they are somehow gifted and different from anyone else, like a chosen one. They feel like there is nobody other than them is suitable for the task.
(2) Development and accomplishment: this is the internal drive of developing skills, making progress and overcoming obstacles. Levels, leaderboards, experience points, skill development, ultimate features unlocked and so on rely on this drive.
(3) Empowerment of creativity and feedback: when a player tries different combinations to figure things out and receives feedback as a consequence of their creativity, the player has the feeling that their decisions created the world, and everything associated with it, even it is virtual.
(4) Ownership and possession: when a player feels like they have possession of something and try to obtain better and better in a never-ending story. Virtual in-game currencies (coins, diamonds, rubies, etc.) are used to accumulate wealth. Likewise, in-game resources in order to achieve goals can be regarded in a similar manner.

Figure 5.1 Yu-kai Chou's Octalysis

Source: https://en.wikipedia.org/wiki/Octalysis#/media/File:Octalysis_Framework.png under a CC BY-SA 4.0 licence.

(5) Social influence and relatedness: this drive incorporates the entire social elements set, including mentorship, acceptance, social responses and companionship, as well as competition and envy. This is where other players take their role. Everyone is their game's hero or antagonist and everyone else are walkers on the scene.
(6) Scarcity and impatience: this drive occurs when the player wants to own something because they cannot have it. This drive causes the player to spend extra money for what they do not have and would like to have without waiting or putting in any effort.
(7) Unpredictability and curiosity: this is the drive that triggers individuals to seek knowledge of what will happen next. Owing to unpredictability, the player's mind is engaged and thinks about what is going to happen. This drive is heavily reliant on chance and randomness. This is the most important drive that keeps the player within the game.
(8) Loss and avoidance: this core drive is based on the avoidance of admitting that everything you did up to this point was useless because you are now quitting. Rather than quitting what they have already acquired, the player continues to play.

Huotari and Hamari (2011) assessed the situation for the service industry and defined gamified service as the sum of the core service and the enhancing service. They used LinkedIn as an example of the core service looking at the user's profile on the website; the enhancing service is the progress bar measuring filling in personal details; the gamified service is the enhancing service's effect on an increase in the perceived value of filling in all the details by invoking a progress-related psychological bias.

When other models on gamification processes are examined, Hunicke *et al.* (2004) underline the different points of views that a player and a designer has. They determine three distinct components (rules, system and fun) and their design counterparts (mechanics, aesthetics and dynamics), respectively. Mechanics is about the algorithm of the game; dynamics is how the mechanics behave as a result of the players' inputs and outputs to the game; and aesthetics refers to emotional responses that a player has during the game.

The last model examined is Werbach and Hunter's (2012) 6D model. The steps predicted in this model are as follows: define business objectives, delineate target behaviour, describe players, devise activity loops, do not forget fun and deploy the appropriate tools.

Gamification Applications

The usage of gamification in education applications (Arnold, 2014; Dicheva *et al.*, 2015; Kiryakova *et al.*, 2014), health (McCallum, 2012; Pereira *et al.*, 2014), as a marketing tool (Hofacker *et al.*, 2016; Moise, 2013) and for tourism marketing (Xu *et al.*, 2013, 2016) are evidenced. Negruşa *et al.* (2015) proposed that gamification contributes to tourism in different ways: as an interface between tourists and local organisations to leverage responsible and ethical behaviour; as a tool leading to collective consciousness for the development of local communities; and as a medium to improve the overall image of a city or location. In their study, they stated that by implementing gamification for marketing purposes, companies in the tourism sector comprehend the use of game elements as an occasion for redesigning loyalty programmes, adding innovation to their promotion campaigns and developing new products. Table 5.2 demonstrates the inward and outward functions of gamification in tourism.

Corrêa and Kitano (2015) analysed the Brazil Quest Game in terms of gamification and evaluated the game as a valid means to diversify the promotion of tourism promotion using technology. Sever *et al.* (2015) proposed that gamification may have some benefits for the online marketing and advertising activities of businesses and destinations. Applications using gamification is funnier and brings more fun, applications that use gamification encourages online engagement, boosts user-generated content, enhances the chance of viral advertising and increases revenue.

Sigala (2015) analysed the contribution of gamification in the purchasing process of tourism products in three phases: before the purchase, during the purchase and after the purchase. The study showed that

Table 5.2 Benefits of gamification in tourism

	When	Benefits	How
Outward functions	Before	• Increase brand awareness, inspiration/dream • Generate interests • Information conversion to sales/ purchase	Tourism-specific games (online games; mobile games) feed tourism information to potential tourists, generate interests, stimulate visiting inspiration, increase brand awareness; virtual reality/3D technology as a technological tool supports these functions
	During	• Engagement/ interact • Enhance experiences • Entertainment tool	Location-based games encourage on site engagement with the destination, augmented reality games interact with the player in real surrounding tourism attractions; gaming as an entertainment tool to kill time
	After	• Loyalty/repeat visit • Share experiences • Recall memories • Advocate/ ambassador	Online games recall memories, encourage sharing experiences, inviting friends, advocate the destination, rewards/coupons in the game and gamified loyalty programmes encourage repeat visit
Inward functions		• Employee management • Employee training • Generate fun interests	Use of games to familiarise the employee, use for staff training

Source: After Xu, Buhalis & Weber (2017)

gamification may be used for crowdsourcing and it affects customer behaviour in all phases of the consumer behaviour process.

Borrero *et al.* (2015) conducted research into a museum with a gamification application that was downloadable onto mobile phones and found out that most users considered that the application not only added a dynamic factor, but also helped them have a more enjoyable and playful visit. Kovačević *et al.* (2014: 21) carried out a study with the Serbia Convention Bureau on usage of gamification in destination marketing. They defined gamification as 'a tool for developing relationship with potential clients and retaining existing clients by using non-typical and dynamical game environment in communicating destination experiences.' This chapter aims to introduce general knowledge about gamification and address the conceptual space of gamification.

Air Canada's Earn Your Wings

Canada's largest airline, Air Canada, was founded in 1937 and is one of the top 10 world's largest passenger airlines by fleet size. Air Canada's Earn Your Wings loyalty programme focuses mainly on badges. It centres on how many wings you gather as a result of an Air Canada connected round trip from Vancouver to Paris via Toronto. The trip consists of standard airports (Vancouver and Toronto) and a destination badge airport (Paris). For each flight, from a standard airport to a standard airport, a user gains 200 wings (100 for take-off and 100 for landing). If the landing or take off takes place at a destination badge airport such as Paris, as in this example, user gains 1500 wings (750 for landing and 750 for take-off).

A user can gain wings (points) from Air Canada's campaign by flying each time with Air Canada, flying in and out of specific airports and completing time limited challenges and collecting badges. There is also an altitude status for users who gain a determined amount of altitude qualifying miles (AQM), altitude qualifying segments (AQS) and altitude qualifying dollars (AQD), which are also obtained with each flight. These statuses are prestige, elite and super elite. Users who have an altitude status may gain access to exclusive services such as, but not limited to, prioritised airport services and lounge access.

The application uses badges to trigger the core drive 'development and accomplishment' and encourages a variety of consuming behaviours by offering numerous badges for different missions. As of June 2019, there are currently 120 different badges in six subgroups including challenges, destination, super destination, participation, globetrotter and cabin. The target audience of the application is current clients who fly with Air Canada. As many badges and wings as the customer receives, they try to obtain another badge or to reach another level, and that is where the loss and avoidance drive plays a part. The customer does not tend to give up what they have already and that leads to an increase in brand loyalty and a tendency to become more engaged.

Turkish Airlines QR (Quick Response) Flags Challenge

Turkish Airlines is the national flag carrier airline of Turkey; it operates scheduled services to nearly 300 destinations in Europe, Asia, Africa and the Americas, making it the fourth largest carrier in the world by number of destinations. In 2012, the Summer Olympics were held in London, UK. Turkish Airlines transformed national flags into QR codes and then placed them on digital bus shelters all over London. Users found the QR code and scanned it with their smartphones automatically to 'check-in' to that flag. The participant with most check-ins won a ticket to one of Turkish Airlines' destinations. The campaign was held by the internationally known advertisement agency, McCann (formerly McCann Erikson).

This campaign clearly targets both current customers, since the challenge gives an option to fly any country with the airlines, and potential customers since the QR flags visually attract anyone who passes by. Time constraints, clear goals and a challenge notion among the game design elements were intensively used in the campaign. The campaign is a good example of a short-term campaign with wider target audience and associated an international event (the Summer Olympics in this case) with the brand in customers' minds. The customers are mainly driven by unpredictability and curiosity as there is no map of QR flags to show where they can be found, and social influence as this challenge may easily turn to a scavenger hunt similar to the mobile game, Pokémon Go.

Marriott My Hotel

Marriott was founded in 1927 as a route beer stand and then expanded to a chain of hotels and restaurants. Marriott's application is called Marriott My Hotel and targets current staff and potential candidates. The game puts users in various parts of a hotel, setting tasks against the clock so that a user can earn points to move on to harder levels or other locations in the property. The application was developed from employee feedback and in order for the company to show that they pay attention to their employees' ideas. The 'game' can be played by other users, as well as the employees of the chain, and in this manner the company also promotes itself globally and thus increases brand awareness.

The app uses game design elements more than any other examples in this chapter. The application's target audience is the current staff in Marriott Hotels and potential staff who want to be recruited within the hotel chain. It is designed to function as an entertainment tool as well as an employee training tool. But things did not go as planned and although it was an innovative gamification attempt, after one year the company decided to remove My Marriott Hotels from Facebook, since it failed to meet the company's original objective in attracting potential employees (Robson *et al.*, 2016).

Starwood's SPG (Starwood Preferred Guest) Programme

Starwood Hotels and Resorts Worldwide, Inc. is an American hotel and leisure company which was founded in 1969 as a real estate investment trust. The company was purchased by Marriott Hotels in 2016. Starwood owns some of the most well-known brands in hospitality, including Sheraton, Westin, St Regis, Le Méridien and so on. After the merger with Marriott Hotels, the Starwood Preferred Guest programme announced it would provide reciprocal benefits with Marriott's own guest reward programme (Trejos, 2016). The app was designed by Stephen Gates.

Starwood's SPG (Starwood Preferred Guest) programme is basically a customer loyalty programme in which customers gain starpoints. They can be redeemed in a number of ways, including free stays in one of the chain's hotels, airline tickets, transferring to airline reward programme points, used as cash for stays, rare opportunities such as a red carpet film premiere and merchandise points with partners on the Amazon website.

Other Gamification Examples

Gamification is a widely used notion in the hospitality and travel industries. It is clearly not possible to review all of the examples here, but it is possible to mention a few. American Airlines has one of the earliest frequent flyer programmes which began in 1982 (Xu *et al.*, 2013). The Miles and More programme by Lufthansa that offers two kinds of miles, that is, 'status miles' and 'award miles', as well as having a level system for different amounts of miles accumulated (Schrape, 2014): level 1 is called frequent-flyer, level 2 is senator status and level 3 HON-circle status. HON-circle status customers have advantages of bonus miles, a meet and greet service in some airports, additional baggage allowance and special treatment from the crew.

Other well-known examples of gamification in the hospitality and airline industries are Delta Airlines' Red Coats Challenge, American Airlines' Advantage Passport Challenge, JetBlue's TrueBlue Badges, App Places by LeClub Accor Hotel and IHG's (InterContinental Hotels Group) Priority Club Rewards.

Discussion and Conclusion

In this chapter, among the current examples to construct a balanced analysis sample, two examples are taken from the travel industry and two examples are taken from the accommodation industry, Air Canada's Earn Your Wings, Turkish Airlines QR Flag Challenge and Marriott International's Marriott My Hotel and Starwood Hotels' SPG programme, respectively. The details of the taken examples are given in Table 5.3.

The criteria used to evaluate and analyse the gamification applications are as follows:

(1) Game design elements used in the application: applications were reviewed in order to find out the game design elements used. This process may be determined as a brief content analysis and it is found that goals, badges, leaderboards and virtual currencies are the most used elements. According to Deterding (2011b), game design mechanics, interface patterns and game design are greatly used, but game models and game design models are rarely seen as they are not as easy to install as the other levels.

(2) Target audience: the target audience of the applications may be either current clients or potential clients and employees. Current and

Table 5.3 Analysis summary of reviewed gamification applications

	Earn Your Wings	**QR Flags Challenge**	**Marriott My Hotel**	**SPG programme**
Sector	Travel (airlines)		Accommodation (hotels)	
Company	Air Canada	Turkish Airlines	Marriott International, Inc	Starwood Hotels and Resorts Worldwide, Inc
Game design elements used	Badges, leaderboards, virtual currencies	Time constraint, clear goals, challenge	Leaderboards, levels, time constraint, limited resources, clear goals	Clear goals, challenge, virtual currencies
Target audience	Current clients	Current and potential clients	Employees	Current clients
Core drives	Loss & avoidance Ownership & possession Development & accomplishment	Development & accomplishment Unpredictability & curiosity Social influence & relatedness	Loss & avoidance Ownership & possession Development & accomplishment	Epic meaning & calling Ownership & possession Development & accomplishment
Time	Long-term	Short-term	Long-term	Long-term
Benefits	Generate fun Interests Engagement/ interact Entertainment tool Increase brand awareness Inspiration/dream	Entertainment tool Engagement/ interact Increase brand awareness Inspiration/ dream	Engagement/ interact Employee training Enhance experiences Entertainment tool	Loyalty/repeat visit Enhance experiences Increase brand awareness Inspiration/dream

Source: After Cilingir Uk and Gultekin (2017)

potential clients are the most targeted audiences but employees were also the target audience of the gamification application in one case.

(3) Duration of the application: the applications are generally designed to be in use for a long time and one case (Turkish Airlines Flag Challenge) was designed to be in use for a short period, as the campaign was designed for a specific event (the 2012 London Olympics). Short-term applications are especially useful if the company is aiming to increase brand awareness by using the density of people in a specific event. These events may be sport events (as it is in this case), festivals, art exhibitions etc. Long-term applications are useful when integrated to ongoing information system applications, such as CRM and database marketing.

(4) Core drives: core drives were derived from Chou's Octalysis. The most common core drive is development and accomplishment where users

believe that they make progress and overcome challenges. Moreover, ownership and possession drive is also widely used through the users' instinct to own as they accumulate wealth within the game by means of virtual currencies (starpoints, coins, AQD). Loss and avoidance is also used in two examples (Earn Your Wings and Marriott My Hotel) to prevent users to quit the 'game'.

(5) Inward and outward benefits of the application: when benefits of the applications were examined, it was found that creating engagement/interaction and using the application as an entertainment tool were common for nearly all of the applications. That situation reflects the fact that the partial use of game elements leads to entertainment and thus motivates individuals to interact and engage. Consequently, this process ends up with an increase in brand awareness.

A new unified model is proposed in Figure 5.2 by determining the gamification as a process and as a process diagram. Game design elements would be inputs as they are the main content that makes an application 'gamified'. Excessive usage of the game design elements has the risk of shifting the conceptual place of the application to a serious game. To maximise the benefits, inward and outward functions have to be evaluated and used separately in general, as functions carried depend heavily on the target audience. An application targeting employees or clients can have benefits as the consequence of inward functions of gamification and an application targeting current or potential clients vice versa.

Core drives should also be carefully selected because excessive usage may cause confusion in users' feelings, ideas and behaviours. Time and

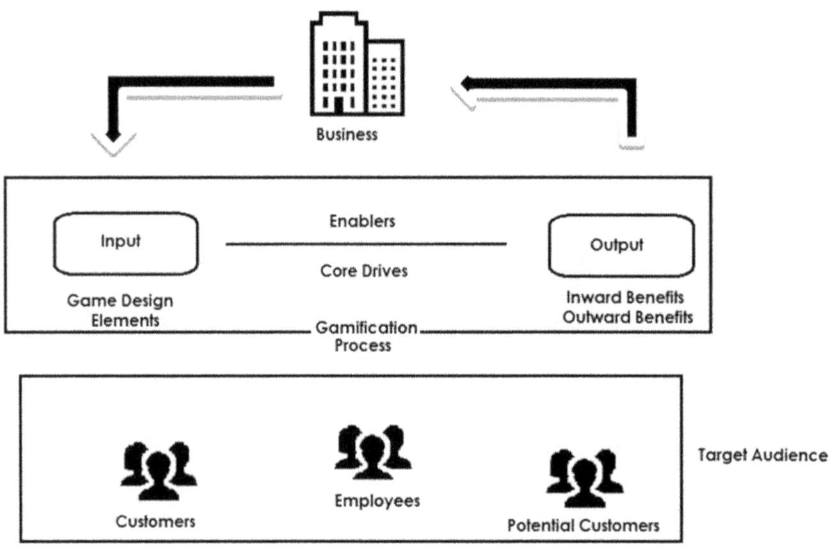

Figure 5.2 Gamification process model

space dimensions of the application are also a component that should be determined. Applications with limited space may only be offered if the application targets a narrow market based on a specific geographical location. Lastly, it is important to get feedback from the target audience as the process should be controlled and revised according to their needs and demands. Social media usage is also crucial to obtain new users.

This chapter critically evaluates the current applications of gamification in the travel and hospitality industries. It proposes a new model where gamification is determined as a process: game design elements as input to the process, benefits of both inwards and outwards functions of gamification as an output of the process, core drives as enablers and the target audience as affected. In future, each subsector of tourism should be examined more precisely as their needs and priorities may differ in terms of gamification. Empirical studies about users' perceptions of gamification applications can improve the design of these applications, and improve their usability and desirability.

Acknowledgement

This chapter is an extended version of the study 'Gamification applications in hospitality and travel industry: Presenting current situation and offering a unified model using new media' written for and presented at the Third International New Media Conference held on 20 April 2017 in Gelisim University, Istanbul.

References

Arnold, B.J. (2014) Gamification in education. *Proceedings of the American Society of Business and Behavioral Sciences* 21 (1), 32–39.

Borrero, F., Muñoz, P.C.S. and González, G.R. (2015) Gamification techniques in tourism, application test, Casa Mosquera Museum. *Sistemas & Telemática* 13 (33), 63–76.

Caillois, R. (2001) *Man, Play, and Games*. Chicago: University of Illinois Press.

Chou, Y. (2015) Octalysis, complete gamification framework. Available at: http://yukaichou.com/gamification-examples/octalysis-complete-gamification-framework/ (accessed 30 January 2017).

Cilingir Uk, Z. and Gultekin, Y. (2017) Gamification applications in hospitality and travel industry: Presenting current situation and offering a unified model using new media. In D. Akcay, T. Sarikaya and O. Dogan (eds) Proceedings of the III International New Media Conference. Istanbul: Istanbul Gelisim.

Corrêa, C. and Kitano, C. (2015) Gamification in tourism: Analysis of Brazil Quest Game. The National Council for Scientific and Technological Development, February. Available at: http://3ws1wk1wkqsk36zmd6ocne81.wpengine.netdna-cdn.com/files/2015/02/SP05_MobileSession_Correa.pdf (accessed 30 January 2017).

Deterding, C.S. (2016) Make-believe in gameful and playful design. In P. Turner and J.T. Harviainen (eds) *Digital Make-Believe* (pp. 101–124). Basel: Springer.

Deterding, S., Khaled, R., Nacke, L. E. and Dixon, D. (2011) Gamification: Toward a definition. In *CHI 2011 Gamification Workshop Proceedings* (Vol. 12), (pp. 12–15), May 7–12, 2011. Canada: Vancouver BC.

Dicheva, D., Dichev, C., Agre, G. and Angelova, G. (2015) Gamification in education: A systematic mapping study. *Journal of Educational Technology & Society* 18 (3), 1–14.

Gaver, W.W., Bowers, J., Boucher, A., Gellersen, H., Pennington, S., Schmidt, A., Steed, A., Villar, N. and Walker, B. (2004) The drift table: Designing for ludic engagement. CHI'04 Extended Abstracts on Human Factors in Computing Systems (pp. 885–900). Vienna, Austria: ACM Press.

Groh, F. (2012) Gamification: State of the art definition and utilization. In N. Asaj *et al.* (eds) Proceedings of the 4th Seminar on Research Trends in Media Informatics (pp. 39–46). Ulm University.

Hamari J. and Eranti, V. (2011) Framework for designing and evaluating game achievements. *Proceedings of DIGRA 2011: Think Design Play*. The Netherlands: Hilversum.

Hamari, J., Koivisto, J. and Sarsa, H. (2014) Does gamification work? A literature review of empirical studies on gamification. 47th Hawaii International Conference on System Sciences (HICSS) (pp. 3025–3034). Hawaii: IEEE.

Hofacker, C.F., De Ruyter, K., Lurie, N.H., Manchanda, P. and Donaldson, J. (2016) Gamification and mobile marketing effectiveness. *Journal of Interactive Marketing* 34, 25–36.

Huizinga, J. (2014) *Homo Ludens*. London: Routledge.

Hunicke, R., LeBlanc, M. and Zubek, R. (2004) MDA: A formal approach to game design and game research. Proceedings of the AAAI Workshop on Challenges in Game AI, 4 (1), pp. 1–5. San Jose, CA: AAAI Press.

Huotari, K. and Hamari, J. (2011) 'Gamification' from the perspective of service marketing. *ACM Conference on Human Factors in Computing Systems (Gamification workshop)*. May 7–12, 2011, Vancouver, BC, Canada.

Kiryakova, G., Angelova, N. and Yordanova, L. (2014, October) Gamification in education. In T. Bulbul and Y. Cakici (eds) Proceedings of the 9th International Balkan Education and Science Conference, 16–18 October 2014 (pp. 679–684) Edirne.

Kovačević, I., Zečević, B. and Veljković, S. (2014) Gamification concept: Theoretical framework and destination marketing management practice. *Ekonomika Preduzeća* 62 (5–6), 315–322.

Malone, T. W. (1981) Toward a theory of intrinsically motivating instruction. *Cognitive Science* 5 (4), 333–369.

Martens, A. and Mueller, W. (2016) Gamification – A structured analysis. In J.M. Spector et al. (eds) IEEE 16th International Conference on Advanced Learning Technologies (ICALT) (pp. 138–142). 25–28 July 2016 Austin: TX.

McCallum, S. (2012) Gamification and serious games for personalized health. *Studies in Health Technology and Informatics* 177, 85–96.

Moise, D. (2013) Gamification – The new game in marketing. *Romanian Journal of Marketing* 2, 29–33.

Negruşa, A.L., Toader, V., Sofică, A., Tutunea, M.F. and Rus, R.V. (2015) Exploring gamification techniques and applications for sustainable tourism. *Sustainability* 7 (8), 11160–11189.

Nicholson, S. (2019) A recipe for meaningful gamification. In L. Wood and T. Reiners (eds) *Gamification in Education and Business* (pp. 1–20). New York: Springer.

Pereira, P., Duarte, E., Rebelo, F. and Noriega, P. (2014) A review of gamification for health-related contexts. In A. Marcus (eds) *International Conference of Design, User Experience, and Usability* (pp. 742–753). Cham: Springer.

Robson, K., Plangger, K., Kietzmann, J.H., McCarthy, I. and Pitt, L. (2016) Game on: Engaging customers and employees through gamification. *Business Horizons* 59 (1), 29–36.

Salen, K., Tekinbaş, K.S. and Zimmerman, E. (2004) *Rules of Play: Game Design Fundamentals*. Cambridge, MA: MIT Press.

Schrape, N. (2014) Gamification and governmentality. In M. Fuchs, S. Fizek and P. Ruffino (eds) *Rethinking Gamification* (pp. 21–45). Lüneburg: Meson Press.

Sever, N.S., Sever, G.N. and Kuhzady, S. (2015) The evaluation of potentials of gamification in tourism marketing communication. *International Journal of Academic Research in Business and Social Sciences* 5 (10), 188–202.

Sicart, M. (2008) Defining game mechanics. Game studies. *International Journal of Computer Game Research*, 8 (2). Available at: http://gamestudies.org/0802/articles/sicart (accessed 30 January 2017).

Sigala, M. (2015) Gamification for crowdsourcing marketing practices: Applications and benefits in tourism. In F.J. Garrigos-Simon, I. Gil-Pechuan and S. Estelles-Miguel (eds) *Advances in Crowdsourcing* (pp. 129–145). Cham: Springer.

Trejos, N. (2016) Marriott, Starwood merger is complete, loyalty programs will reciprocate. Available at: http://www.usatoday.com/story/travel/roadwarriorvoices/2016/09/23/marriott-starwood-merger-complete-and-loyal-customers-get-reciprocal-benefits/90885318 (accessed 21 February 2018).

Werbach, K. and Hunter, D. (2012) *For the Win: How Game Thinking Can Revolutionize Your Business*. Philadelphia, PA: Wharton Digital Press.

Werbach, K. and Hunter, D. (2015) *The Gamification Toolkit: Dynamics, Mechanics, And Components for The Win*. Philadelphia, PA: Wharton Digital Press.

Xu, F., Weber, J. and Buhalis, D. (2013) Gamification in tourism. In Z. Xiang and I. Tussyadiah (eds) *Information and Communication Technologies in Tourism 2014 January* (pp. 525–537). Cham, Dublin: Springer.

Xu, F., Tian, F., Buhalis, D., Weber, J. and Zhang, H. (2016) Tourists as mobile gamers: Gamification for tourism marketing. *Journal of Travel & Tourism Marketing* 33 (8), 1124–1142.

Zichermann, G. and Cunningham, C. (2011) *Gamification by Design: Implementing Game Mechanics in Web and Mobile Apps*. Newton, MA: O'Reilly Media Inc.

6 Innovating the Restaurant Industry: The Gamification of Business Models and Customer Experiences

Marianna Sigala and Elin Nilsson

Introduction

Gamification refers to designing products, services and organisational practices to afford similar experiences to games, with the aim to create value and affect people's behaviour (Huotari & Hamari, 2017). Gamification has been heavily adopted in many sectors including education, retailing, transportation and the military (Johnson, 2016), as well as in social marketing and transformation services such as serious games aiming to change people's lives and behaviours in order to for example moderate alcohol consumption, reduce household energy use and increase physical activity (Mulcahy *et al.*, 2018). Gamification is also incorporated in various contexts and platforms; for example, in websites and e-commerce models (e.g. reputation points systems of Uber, Amazon and Airbnb); branded business apps such as e-banking, sporting (Nike+) and retail mobile apps (Zhao & Balague, 2015). Also, Domino's Pizza Hero app provides a game in which users can knead dough, apply sauce, sprinkle cheese and add ingredients to make a high-quality pizza within a limited time.

Tourism is not an exception to the gamification trend (Sigala, 2015a). Gamification has been examined and applied in various tourism contexts, including the following: motivating and boosting online users' participation and engagement (e.g. through reputation point systems including points, badges and leaderboards (Sigala, 2015b)); crowdsourcing content in online marketing strategies (Sigala, 2015b; Xu *et al.*, 2016); destination apps (e.g. treasure hunt mobile games) aiming to motivate and help tourists to navigate, explore and learn a destination and/or a tourism attraction such as a heritage site, a zoo or a national park (Corrêa & Kitano, 2015); building augmented reality games that enhance the tourists' experience (e.g. Pokémon Go) (Aluri, 2017; Mesároš *et al.*, 2016; Xu *et al.*, 2017); motivating tourists to adopt a more sustainable behaviour (Negrusa *et al.*, 2015); and for gamifying customer relationship management programmes and boosting customer loyalty (Sigala, 2015b). This wide adoption of gamification is responsible for the current transformation of our everyday life, reality and culture (Hamari, 2013).

However, the previous literature has predominately examined the use of gamification as a way for transforming a sole business practice and/or operation into a fun and entertaining activity that can in turn attract consumers' attention and enhance their behavioural and attitudinal behaviours towards the brand and the gamified activity. In contrast to the majority of the previous studies, this chapter examines the use of gamification for building and implementing a whole business model of a company. To achieve that, the chapter examines the gamified app developed by a restaurant chain called Pinchos (a Swedish restaurant chain), which has innovated the restaurant industry. Since its inception, Pinchos used gamification for developing the business concept of its restaurant and the value proposition of the dining experience. In this vein, this chapter expands the gamification literature by analysing how Pinchos used gamification not only for automating and gamifying the restaurant operations (such as food ordering and paying), but primarily for developing its business concept including its value proposition and dining experience.

Gamification: Theory and Practice

Gamification has been defined as 'a process of enhancing a service with affordances for gameful experiences in order to support user's overall value creation' (Huotari & Hamari, 2012: 19). A simple application of gamification is the inclusion of game design elements, such as scoring, points, leaderboards and badges in a non-gaming context (Deterding, 2012). Companies increasingly gamify business practices (such as website use, customer participation in services and creation of online content, loyalty programmes) (Sigala, 2015b), because gamification can achieve the following: increase customer activity such as shopping, contribution of user-generated-content; direct positive user behaviours such as good treatment of Airbnb guests and Uber passengers; entertain customers (Hofacker *et al.*, 2016); facilitate social interactions (Sigala, 2015b); reflect novelty (Johnson, 2016) increase trendiness (Darejeh & Salim, 2016); and create brand awareness, product adoption, strengthen loyalty, deeper engagement and positive word-of mouth (Leclercq *et al.*, 2018). Overall, the positive impacts of gamification on both the behavioural and attitudinal behaviour of consumers is widely documented in the literature (Mulcahy *et al.*, 2018; Sigala, 2015b).

To better understand the application of gamification and how it influences user behaviour, Robson *et al.* (2015) proposed the mechanics–dynamics–emotions framework (MDE), which conceptualises gamification from an input–process-outcome approach. Mechanics include the goals, rules, settings, types of interactions and the boundaries of the situation to be gamified, which are always decided by the game developer. Game dynamics are behaviours and interactions that emerge from customers' gamified experience, and which include both desired behaviour (e.g. users creating more online content) and unintended behaviours (e.g. cheating) (Elverdam & Aarseth, 2007). Effective application of gamification implies that the designer has to develop a funware design that matches the game dynamics with the users' motivations to play the game (Sigala, 2015b). In other words, gamification can bring the desired outcomes of the users' behaviour only when the game dynamics that the users experience while 'playing' the game provide them the benefits and the value that they seek to gain by 'playing' the game. According to Bartle (1996), there are four player types that he labelled as achievers, explorers, killers and socialisers. Thus, game mechanics such as points can effectively work to motivate and boost the performance of achievers but they will not work for explorers and socialisers, who are more attracted and motivated by the game and experience flow through game mechanics such as fantasy, interactions and character-role playing. Finally, emotional components include the positive and negative affective reactions induced by game-like settings (Robson *et al.*, 2015).

The literature identifies numerous game mechanics (Mulcahy *et al.*, 2018; Sigala, 2015b). However, a useful way to categorise game mechanics and to also design gamification applications is their classification into game structure, game tasks and game achievements (Shen & Joppe, 2019).

Game structure offers a background and supports the process of a game. Game designers develop rules that specify how to complete a game, and a storyline that attracts users' attention, curiosity and participation in the game by creating an engaging environment whereby the users feel that they are part of it and they can influence it. game structure elements such as fantasy, progression (e.g. users' movement to advanced game levels increasing the challenge and ensuring continuous flow experience during the game), uncertainty and feedback (e.g. rewards or penalties for players' behaviour showing how to better play the game) are used for pushing the process forward.

Game tasks refer to different gamified activities that the users will be engaging with 'play the game' such as quizzes, challenges and quests, such as testing knowledge, overcoming obstacles, checking in on places and sharing information, which arouse people's interest in playing.

Game achievement refers to accomplishments or benefits of playing a game, including points, badges, levels or status, leaderboards, gifts, rewards and other incentives.

Points are very frequently used in gamification for evaluating the players' performance and motivating user participation. However, although points are common in gamification, it should be noted that effective gamification should not be simply be a pointification exercise. This is because points (as previously explained) may not be an effective game mechanic to appeal and motivate all types of 'gamers', while it can also result in negative user experiences and outcomes, such as, burn-out, demotivation and user long-term disengagement (Basten, 2017; Huotari & Hamari, 2012; Sigala, 2015b). This is because when gamification stops providing the points and/or when the users have accumulated enough points, they stop being motivated while they are disengaged. On the contrary, gamification should be more than pointification by aiming to motivate user engagement by providing and blending both intrinsic (e.g. fun, entertainment, escapism benefits) and extrinsic motivation (e.g. points, incentives) (Hamari, 2017). Gamification tries to transform activities that require external motivation for activities that are fun and therefore motivating in themselves. The combination of fun and reward is an effective way of encouraging behaviour (Zichermann & Cunningham, 2011) and therefore the affordances of gamification to motivate and foster people's engagement and participation into the gamified tasks. The biological explanation of the behavioural impacts of gamification are also discussed in the literature (Huotari & Hamari, 2012): collaboration between challenge, achievement and reward triggers dopamine in the brain, which actually strengthens the biologically positive effect that make people feel joy and commitment.

Customers increasingly demand to actively participate and design their own services and experiences, while modern technologies and service design empower customers to become active co-creators of their service experiences (Ballantyne & Nilsson, 2017). However, for value to be co-created, customers should be skilled, ready and motivated to engage and participate in the co-creation process as well as interact with other actors for exchanging and integrating resources (Grönroos, 2011; Grönroos & Voima, 2013; Ramaswamy, 2011). Payne *et al.* (2008) argue that customers will be co-creators of value when they experience motivation and a willingness to be involved in the service. Because of its affordances to motivate and foster customer participation, gamification is a useful technique for ensuring the successful implementation and outcomes of co-creation processes requiring the active engagement of customers.

Methodology

The study aimed to provide a holistic perspective about the design and the impacts of the gamified app of Pinchos. To that end, the study used a variety of methods for collecting data from various restaurant stakeholders involved in the design and implementation of the gamified app.

During 2017 and 2018, the researchers visited and experienced a meal at various Pinchos restaurants located in three different cities in Sweden. Observational data and personal experiences enabled the researchers to obtain useful information and a first-hand understanding about how the design of the gamified app influences the design of the restaurant servicescape, the customer-to-staff interactions and the restaurant operations/processes (e.g. ordering, payments etc.).

In-depth interviews were also conducted with the marketing communicator/brand manager of the restaurant chain (Caisa Lintunen) in spring 2017 and 2018. The interviewee provided particularly useful insights about Pinchos' approach to developing the gamified app for developing its business model, supporting their digital interactions with their guests and building customer engagement on social media for promoting the restaurant. The researchers also collected additional information about Pinchos' digital interactions with guests by reading and analysing the restaurant's Facebook page and online guests' discussions.

Pinchos: Company Background

Pinchos (http://www.pinchos.se) was founded in 2010 by Magnus Larsson and Fredrik Mattsson, and thanks to the digitisation and gamification of its business model, the company has experienced exponential growth rates. The founders had promised each other to never get into the restaurant business again. They had tried several times and always came back to the same conclusion – it was too hard and the profitability too low (if it even existed). But when digitalisation started to take off, they discussed how the restaurant industry could benefit from this. Their first idea was to build a digital system that would fit restaurant operations and then sell it off. However, as it was difficult to find a buyer for the system, they opened a 'try-out restaurant' in 2010 in Smögen to prove that the system could effectively work. Unfortunately, the system did not work as well as they had planned and thought. The consumers had troubles using the app but they kept the project running by polishing the digital system and making it more user friendly. In order to really test the system, they bought a permanent restaurant. In 2012, the first Pinchos restaurant concept opened in Gothenburg, which is distinguished by being the world's first and only restaurant where all orders are made via the chain's proprietary and developed mobile app. Since then, Pinchos has become known as one of the pioneers of integrating apps in the restaurant industry.

In 2014, Pinchos decided to use a franchise to expand its business in Sweden and abroad. Pinchos is Sweden's fastest growing restaurant chain with 63 current restaurants and another six restaurants being planned: 61 restaurants are located in Sweden, two in Norway, one in Denmark, one in Finland and one in Germany; five restaurants are planned to open in five more locations in Sweden.

Pinchos has also expanded and grown its technological business in order to also address its operational needs. In October 2017, Pinchos acquired the advertising agency 'How Do You Do' in order to acquire and build the required capabilities for addressing the increased communication needs, especially when expanding the restaurant concept beyond the Swedish borders.

The Pinchos Mobile App

Since 2012, when the first Pinchos restaurant opened, public media criticised the use of a mobile app in a restaurant setting. This technology was rather very new in Sweden and there were many sceptics. However, Pinchos developed and extracted the benefits of this app before anyone else in the restaurant industry could vision and predict its benefits. Caisa Lintunen (brand manager at Pinchos) stated that Pinchos had been working hard on making the app as simple as possible so that it was easy to use and this was also why it had been such a great success.[1] Since their arrival in 2008, smartphones have been heavily adopted in Sweden and many companies have started using mobile apps as well. However, their apps have not been as successful as the Pinchos app, simply because their customers are offered the option not to use them. If customers wish to have a meal at Pinchos, they have to use the Pinchos app (Figure 6.1), because the mobile

Figure 6.1 Pinchos' mobile app

app was used to build its business model by embedding gamification into its value proposition and dining experience. According to Lintunen, the lack of an opt-out option is one of the success factors of the Pinchos app.

The customers handle table booking, menu browsing, ordering and payment via the app on their own mobile phone (Figure 6.2). In the early version of the app, customers entered their card details, which were then saved to enable direct payments for restaurant visits at the chain. Even though the system worked well, Pinchos' pay feature changed in December 2017 to a version that is even safer. The upgrade meant that all card data previously saved was deleted and replaced by a system that is linked to BankID. Payment can now be made directly in the app by card or by invoice at no extra charge. Pinchos' partner for the payment solution is the Swedish payment institution Paynova, one premium provider of online trading solutions that are constantly developing new fraud protection.

Since its launch in 2012, Pinchos has been Sweden's first and only app restaurant. The success of its app is clearly evident in the numbers reflecting the distribution and the usage of the app. The app has been downloaded more than 1.5 million times (until 2017), and 100% of customers use it for ordering and 75% for payments. Pinchos has received several awards for its app, including Framtidens Franchise (2016, Svenska Franchise), Årets Digitala Marknadsföring (2017, Svensk Franchise) and EY Årets entreprenör i Västsverige (2017, EY).

The functionality of the app has been expanded to also include a loyalty programme. Customers gain loyalty and bonus points when ordering food, paying their restaurant bill and when interacting and sharing the app's content and activities on social media. This gamification of the loyalty programme of the Pinchos app adds to the gamified restaurant experience that the customers can have and it further builds and motivates customer engagement, spending and loyalty.

The Funware of the Pinchos App: Gamifying the Dining Experience and the Business Model

The dining experience has been conceptualised in various approaches depending on the study goals of the researchers (e.g. Dixon *et al.*, 2009; Namkung & Jang, 2010; Wijaya *et al.*, 2013). In general, the dining experiences are analysed in the following three stages: pre-dining stage; in-dining stage; and post-dining stage. Various studies have shown how technology can improve restaurant operations, including the menu, ordering, booking, online usage, payment-based service innovations and queuing (Dixon *et al.*, 2009). In addition, mobile applications and gamification can be used to accompany solo diners (Brown *et al.*, 2020). However, there is no study so far examining how gamification can be used not only to improve restaurant operations, but also to enhance the customer experience, innovate the dining experience and boost customer loyalty.

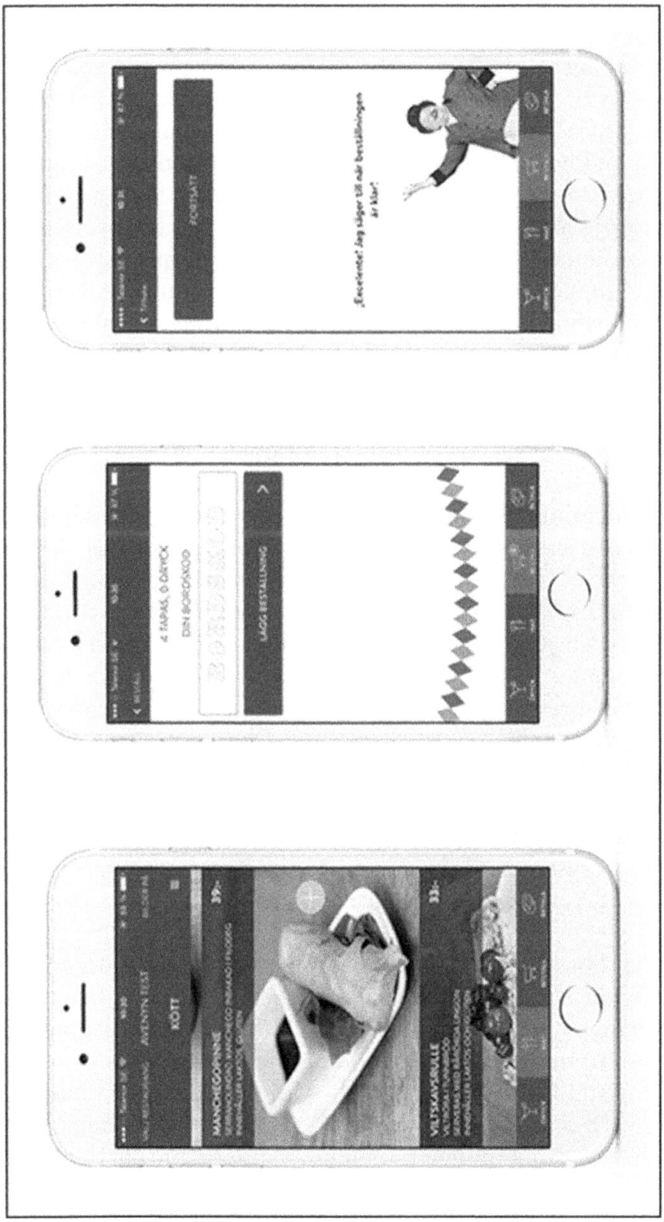

Figure 6.2 Pinchos' mobile app functionality: table booking, ordering and payment

The following will explain how Pinchos has gamified not only the dining experience in all three stages, but their whole business model including the value proposition and operations by using game mechanics that convert a functional, routine and labour demanding function into a 'play' task that both staff and customers are motivated to perform and engage with.

Before the meal experience, the customer can use the app in order to find locations of restaurants, check and book table availability, invite friends for a meal and check the menu. When the customers arrive, they are greeted and hosted by the restaurant staff. The menu is found in the app and the customers can easily see pictures of the food as well as food allergy-information. Pinchos' menu consists of tapas format courses from kitchens throughout the world, such as Spanish quesadillas, Chinese dumplings, Hungarian plank steak, French crème brûlée, American mini-burgers, English fish and chips and Japanese salmon sashimi. Pinchos' goal is to offer everyone something – whatever their taste, lifestyle and wallet size. Just like the food, the drinks menu is inspired by both the customers and people they have met around the world, containing something for everyone; everything from traditional mojitos, sangria, ice-cold beer and good wines to their own playful creations that are in all the colours of the rainbow. The restaurant concept is to offer something for everyone no matter their taste preference or monetary budget. After deciding what to eat, customers need to press order and the personal table code. When the food or drinks are ready, the mobile app makes a circus drum sound alerting customers to collect their food or beverage. The customers can use the app for re-ordering more food and/or drinks and paying their bill when they are ready to go. During and/or even after the meal, the customers can use the app to share their meal experience within their social networks by using social media platforms such as Facebook. The mobile app has gamified all the tasks that customers have to do before, during and after the dining experience.

The bonus system and the points are the major game mechanics used for measuring, rewarding and motivating/boosting the performance of customers when implementing the gamified meal experience tasks (Figure 6.3). For example, whenever someone orders food and/or pays with the app, he/she gets points that can be exchanged for free tapas. The points are a major way for motivating the customers to order more and/or visit the restaurant again for redeeming the points. This is in line with Zichermann and Cunningham (2011) who argue that a combination of fun and reward is an effective way of encouraging behaviour.

Pinchos has also gamified the online marketing tasks that customers can do through the mobile app. Customers are rewarded for recruiting and generating customers. Each time a customer invites friends, they earn points. When this friend visits the restaurant and pays for their meal, the consumer who shared the information with this friend gets 200 points and the friend gets 50 welcome points.

Figure 6.3 Pinchos' mobile app: bonus system

By using the mobile app, the customers are co-creating their dining experience while also having fun and/or being rewarded for their behaviour (e.g. collecting points by ordering and paying for food). This works particularly well for solo diners (Brown *et al.*, 2020). Gamification does blur the distinction between work and play, which is often called 'playbour' (Hamari, 2017).

Storytelling and rules are the core elements of the structure of the Pinchos gamified mobile app. The basic rule is that if someone wants a meal at the restaurant, the only way to book a table, order and pay for food is through the mobile app. Other restaurant rules that the customers need to know in order to 'play the game' and to have a memorable and pleasant meal experience are outlined below. The customers can save up to 1200 bonus points or six free tapas. This is an example of game achievements (Shen & Joppe, 2019), where the customer gets the benefits of 'playing the game'/visiting the restaurant. If the customers choose to use their bonus points as payment, all of the points are deducted when paying. The following conversion rules are used for converting points into tapas:

- 300 points = 1 free tapas
- 600 points = 2 free tapas
- 900 points = 3 free tapas
- 1200 points = 6 free tapas

Pinchos is also planning to add more spontaneous treats for frequent guests and to include special rewards in order to treat loyal customers on their birthdays. This 'surprise' element in customer rewards can further boost their motivation and engagement with the gamification application and boost their brand loyalty and spending in the restaurant. It might also motivate the other player types presented by Bartle (1996), who are not drawn to achievements.

Storytelling has also been used for developing the business model and concept of Pinchos, where the story of Señorita Pinchos is important. The story is about a young woman, Señorita Pinchos, who, following her father's wishes, went out into the world to experience distant cultures and kitchens. She joined the crackling, bustling collection of odd and fabulous characters in a circus.

- I cook the food! Señorita Pinchos stated when she saw the circus appear at the train station.
- Excellent! Said the circus director and helped Pinchos aboard.

Every night after each performance, the circus crew gathered to enjoy the delicious snacks that Señorita Pinchos cooked. They laughed, sang and shared the food enthusiastically, which was a blend of different cultures. Everyone found something that made them feel like eating with friends and family at home. Together, the odd figures formed a circus family in which the nightly feasts constituted the hub. The small dishes were named Pinchos – fun sharing appetizers. The memory of the circus family and Señorita Pinchos still live on in the restaurant.

The mobile app communicates the business concept and the story of the restaurant, which basically informs the customers and creates their expectations in terms of the meal experience that they can have at Pinchos. Thus, the gamified meal experience at Pinchos creates an engaging environment that motivates people to select Pinchos for having a meal and whereby the customers feel part of the story (i.e. the meal experience), which they can influence by participating in the game (i.e. the gamified app). Apart from the mobile app, Pinchos has used the same story for gamifying all its restaurant operations and elements affecting the meal experience, that is the servicescape, the restaurant atmospherics, the role of the staff and their interactions with customers. The staff are dressed as circus conductors and the smell of popcorn, the music in the toilets and the red circus interior are important elements of the restaurant business concept and found everywhere in every Pinchos restaurant. Despite the automation and self-service offered through the mobile app, Pinchos has not cut down on staff numbers. On the contrary, the restaurant employs more staff whose role is to interact with customers, spend time with them and ensure that they have a good dining experience. The waiters are responsible for greeting customers at the door, bringing them to their table, serving them with popcorn and explaining how the app, if they have never visited the restaurant before. The Pinchos' members of staff are trained on how to help customers with the technical parts, so they could be seen as technical support rather than waiters. Caisa Lintunen (brand manager at Pinchos) stated that: 'Usually older customers are more skeptical but after they have downloaded the app and used it, they get a positive surprise. It is really fun to see how proud and positive they are to have managed it.'[2]

Pinchos' staff are instructed to interact with customers not only face-to-face but also online on social media. As much of the dining experience is 'co-created' by the customer online and through the gamified app, which is linked with the social media, the staff also need to be online and interact with customers. Lintunen stresses that Pinchos want to create a closeness to their customers and therefore it is important that the staff answer online customer enquiries quickly, as well as it adding a personal touch to digital communications. Members of staff use emojis, hugs, smiles and hearts a lot in their online communications. In one post on Facebook, Pinchos asked their followers what they wanted to see on the menu in the future:

Customer X: Bruschetta would be a nice addition.

Pinchos: -Looove Bruschetta! I'll tell the chefs this with a big megaphone. <3 Thank you!

Customer Y: Ceasarsallad!!.

Pinchos: A little bird whispered in my ear that it might show up when it gets a little warmer ☺.

Customer Z: -Västerbotten pie with cream and cream fraiche and chives, a plate with sausage, cheese, olives, pickles, aioli etc,.

Pinchos: Hallelujah! You and I would be best friends! The mini-plate & cheese tray is a must! Know that our chefs are working on it 'as-we-speak' ☺ <3

Creative, engaging, informative and edutainment content is also posted online by Pinchos in order to engage and interact with their customers as well as add some educational value to their meal experience. For example, in order to further raise awareness of breast cancer, Pinchos produced an informational film explaining how woman can examine themselves and their breasts in order to detect possible changes at an early stage. The film was posted on Pinchos' social media channels and received a high level of customer engagement; on the restaurant's Facebook page, the film reached over 200,000 users and over 60,000 users interacted with it by sharing and commenting. During October 2017, Pinchos also changed its most popular dessert to a pink version to make a clear connection with the Breast Cancer Foundation. Swedish Caterer was responsible for the raw material costs. The amount for each dessert sold went directly to the Breast Cancer Foundation. When the revenues were counted, it was clear that Pinchos and Svensk Cater received SEK463,767. The entire amount was donated to the Breast Cancer Foundation. This blurring of the offline restaurant meal experience (i.e. dessert design, consumption of the pink dessert) with the online gamified task (ordering and paying for the pink dessert) is a very good example of the use of gamification for social marketing (i.e. raising awareness of a health issue) as well as transformational experiences such as training customers how to self-examine themselves as well as encouraging them to get into the habit and routine of doing it.

Lintunen argued that the digital technology did not make the dining experience antisocial. At Pinchos, the staff members are instructed to spend more time talking to the customers, as their role is not to take orders and payments. The staff members are also trained and directed to create many social meetings and 'social moments' with customers when food and drink are collected. Lintunen stressed that even though Pinchos restaurants might have one waiter less than similar sized restaurants, Pinchos does not aim to use the mobile app to save labour costs. Rather, Pinchos' philosophy is that the technology is used in order to take care of the process and manage the food/drinks logistics and functions, and hence free up time and to enable the staff to take care of the customers. In fact, Lintunen claimed that 'our staff maybe need to work a little harder compared to other restaurants to make sure that they interact with the customers.'

According to Hamari (2017) gamification is all about motivating, training and directing customers' behaviour to co-create value. Co-creation opportunities are very evident in the design of the Pinchos mobile app whereby the dining experience and the restaurant business model is all about customers co-creating their meal experience at every stage by being the co-designers of their meal (food-drinks selection and pairing as well as influencing new dish development), the co-producers of the dining operations (table booking, meal ordering and payments, staff-guests, guests-to-guests interactions) and the co-marketers of the meal experience (reviews, online discussions, sharing of online content) collaborating on the platform. In addition to that, the mobile app allows the restaurant to collect a great deal of information about its guests, such as the popularity of dishes across demographics, locations, busy and slow hours and consumers' habits, which in turn is used by the restaurant in order to further improve its operations. For example, Pinchos have used consumer input from social media to change things in the menu. The candy drinks were the first thing they expanded due to huge interest on social media. Also, more vegan options were added to the menu after questions about vegan food.

Pinchos used gamification to change their restaurant business model in various ways. The bonus system motivates the guests to want to order more food and drinks and to pay with the app in order to receive free tapas in the future and/or accumulate points and exchange them for free services. The mobile app also makes a 'fun sound' when the food and drinks are ready, and which represents a major servicescape element creating the circus environment. Free tapas are provided as rewards for game tasks, which gives the customers a sense of accomplishment/achievement. Customers are able to suggest food and drinks for the future, which gives them a meaning and purpose to using the mobile app. Regarding empowerment, the mobile app is also used for listening to customers: the customer gets feedback when posting something on social media and Pinchos are also posting customers' pictures on their site, tapping into both empowerment and scarcity. Customers help others with their bonus points by inviting each other for a meal, which in turn creates a social networking and co-creation aspect of the system (i.e. supporting viral marketing techniques).

Table 6.1 summarises the funware design of the Pinchos' mobile app by showing how the three elements of game mechanics (game structure, tasks and achievements) are applied in the restaurant business model and operations.

Conclusions and Implications

Gamification is widely adopted for gamifying business operations, converting functional tasks into a fun and entertainment play task in which customers will be motivated and fully engaged to participate in. Although gamification is heavily adopted in many contexts and platforms for gaining the positive behavioural and attitudinal change in consumer behaviour, no research has yet examined the use of gamification for transforming and building the business model and concept of a company. This chapter addresses the gap by analysing the use of gamification by Pinchos (a Swedish restaurant chain) that used gamification for innovating the dining experience and the restaurant industry. The chapter describes how gamification is embedded in, and has influenced every aspect of, the restaurant business model, including the servicescape, staff role, menu design and customer interactions and roles.

As summarised in Table 6.1, the restaurant app empowers guests to actively participate in and co-create their meal experience (before, during and after the restaurant visit), while it also uses gamification for motivating and fostering customer engagement. The Pinchos gamified app has transformed the co-creation activities of the customers and the staff tasks into playbour. As limited research exists about the long-term impacts of gamification and specifically of playbour, it is suggested that future studies are directed to examining the effective design and consequences of this concept in various contexts such as collaborative economy (e.g. Airbnb, Uber), online retailing (e.g. Amazon) and offline contexts.

As technologies are having a profound impact on customer experiences, understanding how to use technology applications for enhancing customer engagement is an important part of a successful digital strategy. Moreover, as the sector evolves and digitalisation plays a bigger part in

Table 6.1 The gamification of the meal experience: The funware design of Pinchos' mobile app

Stage of meal experience	**Game tasks: Customer co-creation**	**Game achievements: Bonus system – points – rewards**	**Game structure: Rules and storytelling**
Before the meal	• Find restaurant locations. • Book a restaurant/table. • Invite friends for a meal. • Check menu – pre-order food and drinks.	• Points when ordering food.	Rules • The mobile app: the only way to plan, have and pay for a meal experience. • Number of points that guests are able to accumulate and save. • Number of points required to order each food and drink item.
During the meal	• Order food. • Re-ordering food and drinks. • Share online content of the meal experience.	• Points when ordering food. • Points when paying for food. • Points when recruiting and generating a guest by promoting restaurant to friends. • Points for sharing online content.	Story-telling Señorita Pinchos' story • The story is about a young woman, Señorita Pinchos, who following her father's wishes, went out into the world to experience distant cultures and kitchens. Online interactions promote the various international dishes that this lady has accumulated and experienced to the guests.
After the meal	• Share online content of the meal experience. • Write – share an online review. • Promote restaurant to friends – recruit a friend as a restaurant guest.	• Points for sharing online content. • Points when recruiting and generating a guest by promoting restaurant to friends. • Special rewards in order to treat loyal customers on their birthdays.	The storytelling design of the servicescape (theatre stage design) • Staff dressing as circus conductors. • Staff's online and offline interactions with guests. • The smell of popcorn. • The music in the toilets. • The red circus interior. • The fun sound of the mobile app when ordering food and drinks. The breast cancer story-telling campaign • Aim to inform guests about breast cancer and activate their support by using customers' spending to support a foundation. • Servicescape design of pink drinks. • Online interactions informing guests about breast cancer. • A campaign with a social cause.

restaurants and cafés, managers need to be prepared to create a new combination of human service and digital efficiency and interaction. The role of the waiters and other restaurant staff will change and managers need to be prepared for that. Pinchos is not the first or only restaurant or café that has implemented an app, but it is still the only restaurant whose mobile app is the sole way in which the guests can plan and have their meal experience (e.g. book a table, make and pay for an order) and whereby the guests have to use their own devices instead of being given a company device. This also adds to the customer loyalty programme and it is easy for the customers to understand the benefits of revisiting the restaurant. When paying at the end of the restaurant visit, the app automatically informs the customer about how many free tapas they are entitled at their next visit; this adds a positive element at the end of the restaurant visit and does not end the experience with a simple payment. In other words, the gamified loyalty programme enriches the customer experience. The gamified app also helps the restaurant to develop and maintain a relation with the customers even after they have left the restaurant or even well in advance of their visit to the restaurant. The gamified app enables the restaurant to digitally interact and engage with the customers by creating, sharing and distributing informative and edutainment content. Overall, the business benefits of the mobile app are evidenced by the fact that Pinchos had been having fully booked restaurants since their opening, even in new cities and locations where a previous customer restaurant experience did not exist.

The way in which Pinchos developed its mobile app and integrated it into its business model also raises several questions for future research. Research is required to better understand the market appeal and impact of a technology application offering no opt-in and/or opt-out option to customers. How do people react to such digital supported services where online co-creation activities demand customers be co-creators of their service and if this is the only option to consume the service? How do different customers react, similarly and/or differently to such service contexts? How does such a gamified service platform supporting customer co-creation influence the customers' perceptions about price/quality and value-for-money propositions, as well as their service quality expectations and evaluations? Research is also required to investigate consumers' perceptions and reactions in bringing and using their own resources (i.e. time and digital devices) for co-producing their meal experience. For example, do customers perceive and expect that the greater the use of their own resources and tools should result in lower service prices and/or quality levels? Obviously, the Pinchos' gamified app provides not only an interesting case study to understand the use and implications of gamification into a service setting, but it also sets the context for investigating whether this mobile app creates a new service paradigm transforming business models and forming new markets and customer perceptions.

Notes

(1) Personal communication, Caisa Lintunen, brand manager, Pinchos.
(2) Personal communication, Caisa Lintunen, brand manager, Pinchos.

References

Aluri, A. (2017) Mobile augmented reality (MAR) game as a travel guide: Insights from Pokémon GO. *Journal of Hospitality and Tourism Technology* 8 (1), 55–72.

Ballantyne, D. and Nilsson, E. (2017) All that is solid melts into air: The servicescape in digital service-space. *Journal of Services Marketing* 31 (3), 226–235.

Bartle, R.A. (1996) Hearts, clubs, diamonds, spades: Players who suit MUDs. Available at: https://www.researchgate.net/publication/247190693_Hearts_clubs_diamonds_spades_Players_who_suit_MUDs (accessed 4 February 2019).

Basten, D. (2017) Gamification. *IEEE Software* 34 (5), 76–81.

Brown, L., Buhalis, D. and Beer, S. (2020) Dining alone: Improving the experience of solo restaurant goers. *International Journal of Contemporary Hospitality Management* 32 (3), 1347–1365.

Corrêa, C. and Kitano, C. (2015) Gamification in tourism: Analysis of Brazil Quest Game. *e-Review of Tourism Research* 6 (1), 1–5.

Darejeh, A. and Salim, S.S. (2016) Gamification solutions to enhance software user engagement – A systematic review. *International Journal of Human–Computer Interaction* 32 (8), 613–642.

Deterding, S. (2012) Gamification: Designing for motivation. *Interactions* 19 (4), 14–17.

Dixon, M., Kimes, S. and Verma, R. (2009) Customer preferences for restaurant technology innovations. Center for Hospitality Research Publications. Available at: http://scholarship.sha.cornell.edu/chrpubs/34 (accessed 12 April 2019).

Elverdam, C. and Aarseth, E. (2007) Game classification and game design: Construction through critical analysis. *Games and Culture* 2 (1), 3–22.

Grönroos, C. (2011) Value co-creation in service logic: A critical analysis. *Marketing Theory* 11 (3), 279–301.

Grönroos, C. and Voima, P. (2013) Critical service logic: Making sense of value creation and co-creation. *Journal of the Academy of Marketing Science* 41 (2), 133–150.

Hamari, J., (2013) Transforming homo economicus into homo ludens: A field experiment on gamification in a utilitarian peer-to-peer trading service. *Electronic Commerce Research and Applications* 12 (4), 236–245.

Hamari, J. (2017) Do badges increase user activity? A field experiment on the effects of recommendations. *Business Horizons* 58 (3), 305–315.

Hofacker, C.F., De Ruyter, K., Lurie, N.H., Manchanda, P. and Donaldson, J. (2016) Game-based mechanic's ability to increase attraction, engagement, and retention in gamification and mobile marketing effectiveness. *Journal of Interactive Marketing* 34, 25–36.

Huotari, K. and Hamari, J. (2012) Defining gamification: A service marketing perspective. Paper presented at the 16th International Academic MindTrek Conference, October, Tampere, Finland.

Huotari, K. and Hamari, J. (2017) A definition for gamification: Anchoring gamification in the service marketing literature. *Electronic Markets* 27 (1), 21–31.

Johnson, C.P. (2016) Gamification in adventure and wilderness sports: A literature review of gamified smart tourism applications: An integrative approach. *Sustainability* 9 (12), 21–62.

Leclercq, T., Hammedi, W. and Poncin, I. (2018) The boundaries of gamification for engaging customers: Effects of losing a contest in online co-creation communities. *Journal of Interactive Marketing* 44, 82–101.

Mesároš, P., Mandičák, T., Mesárošová, A., Hernandez, M.F., Kršák, B., Sidor, C. and Delina, R. (2016) Use of augmented reality and gamification techniques in tourism. *e-Review of Tourism Research* 13 (1/2), 366–381.

Mulcahy, R.F., Russell-Bennett, R., Zainuddin, N. and Kuhn, K.A. (2018) Designing gamified transformative and social marketing services: An investigation of serious m-games. *Journal of Service Theory and Practice* 28 (1), 26–51.

Namkung, Y. and Jang, S. (2010) Service failures in restaurants: Which stage of service failure is the most critical? *Cornell Hospitality Quarterly* 51 (3), 323–343.

Negrusa, A.L., Toader, V. and Sofica, A. (2015) Exploring gamification techniques and applications for sustainable tourism. *Sustainability* 7 (8), 11160–11189.

Payne, A.F., Storbacka, K. and Frow, P. (2008) Managing the co-creation of value. *Journal of the Academy of Marketing Science* 36 (1), 83–96.

Ramaswamy, V. (2011) It's about human experiences...and beyond, to co-creation, *Industrial Marketing Management* 40 (2), 195–196.

Robson, K., Plangger, K., Kietzmann, J.H., McCarthy, I. and Pitt, L. (2015) Is it all a game? Understanding the principles of gamification. *Business Horizons* 58 (4), 411–420.

Shen, Y. and Joppe, M. (2019) Gamification in tourism advertising: Game mechanics and practices. Available at: https://www.researchgate.net/profile/Ye_Shen19/publication/341220120_Gamification_in_Tourism_Advertising_Game_Mechanics_and_Practices/links/5eb47a3ca6fdcc1f1dc81696/Gamification-in-Tourism-Advertising-Game-Mechanics-and-Practices.pdf (accessed 17 December 2020).

Sigala, M. (2015a) Gamification for crowdsourcing marketing practices: Applications and benefits in tourism. In F.J. Garrigos Simon, Y. Narangajavana and J.L. Galdón-Salvador (eds) *Crowdsourcing as a Competitive Advantage for New Business Models* (pp. 129–145). Cham: Springer Verlag.

Sigala, M. (2015b) The application and impact of gamification on trip planning and experiences: The case of TripAdvisor's funware. *Electronic Markets: The International Journal of Networked Markets* 25 (3), 189–209.

Wijaya, S., King, B., Nguyen, T-H. and Morrison, A. (2013) International visitor dining experiences: A conceptual framework. *Journal of Hospitality and Tourism Management* 20 (Supplement C), 34–42.

Xu, F., Tian, F., Buhalis, D., Weber, J. and Zhang, H. (2016) Tourists as mobile gamers: Gamification for tourism marketing. *Journal of Travel and Tourism Marketing* 33 (8), 1124–1142.

Xu, F., Buhalis, D. and Weber, J. (2017) Serious games and the gamification of tourism. *Tourism Management* 60, 244–256.

Zhao, Z. and Balague, C. (2015) Designing branded mobile apps: Fundamentals and gamification. *Computers in Human Behavior* 71, 469–478.

Zichermann, G. and Cunningham, C. (2011) *Gamification by Design: Implementing Game Mechanics in Web and Mobile Apps*. Newton, MA: O'Reilly Media, Inc.

7 Destination Marketing via Gamification: A Case Study of the Austria Adventure Game

Feifei Xu and Shaojin Li

Introduction

As a new type of marketing tool, gamification has been applied into many different fields, such as education, health care, medical and military services (Zichermann & Linder, 2010). It is different from traditional models of marketing, in that gamification emphasises players' enjoyment of immersive experiences, aiming to increase brand awareness and brand loyalty in the game through active engagement and the involvement of consumers (Xu *et al.*, 2014).

The tourism industry is an experience economy. Pine and Gilmore (1999) indicate that active engagement of tourists at various levels (emotional, physical, intellectual and spiritual) contribute to a more meaningful and personal experience (Jennings *et al.*, 2009). Several researchers indicate that gamification is be a popular trend in tourism marketing (Chaffery & Ellis-Chadwick, 2012; Jobber, 2009; Ye & Joppe, 2020). Middleton *et al.* (2009) mention that experiential marketing can generate tourists' interest. Gamification can enhance tourists' on-site experiences by adding fantasy and fun through technologies such as virtual reality, augmented reality and human–computer interaction. This creates an immersive environment for the tourists and increases user engagement, which contributes to the formation of positive attitudes towards the brand (Xu *et al.*, 2016).

Gamification applications in the tourism industry are emerging rapidly; for example, the Smile Land Thailand game developed by Visit Thailand (Xu *et al.*, 2017); RexPlayer, an augmented reality game developed by Regensburg, Germany (Waltz & Ballagas, 2007); the Brazil Quest game developed by Brazil (Corrêa & Kitano, 2015); the Whaiwhai Travel

Book app; a series of Discovery Hong Kong games developed by the Hong Kong Tourist Board (Stadler & Bilgram, 2016) and so on. While academic research on tourism gamification is still lacking, there is emerging academic attention on this topic (Bulencea & Egger, 2015; Corrêa & Kitano, 2015; Negruşa *et al.*, 2015; Sigala, 2015a, 2015b; Xu *et al.*, 2014). But how exactly gamification could impact tourists' decision-making and buying behaviour is unclear and empirical research in particular is particularly limited.

This chapter fills in the gap through conducting empirical research with a purposely developed Austria Adventure Game, which was aimed at potential tourists. Using a quantitative survey, the aim of the research was to examine how exactly gamification could influence tourism marketing. By doing so, the objectives of this chapter were: (1) what are the motivations for playing the Austria Adventure Game? (2) Does gamification increase players' brand awareness? (3) Does gamification influence tourists' intentions to visit the destination?

Following the introduction to gaming in tourism, the chapter starts with a review of the literature on the concept of games, gaming motivation, marketing and branding, and the gamification of tourism marketing. The methodology section explains the questionnaire design, sampling and data collection. The chapter then analyses the results of tourism gamification in the Austria Adventure Game. Finally, the chapter ends with a summary of the findings and suggests areas for future research.

Literature Review

The concept of gamification

While different terms have been used, such as 'serious games', 'game-based learning' and 'pervasive games', there has been no universal definition of gamification (Bogost, 2007; Epstein, 2013; Kapp, 2012; Seaborn & Fels, 2015). However, Deterding *et al.*'s (2011) broad definition, which refers to the use of game design in fields other than pure entertainment, is widely accepted. Uskov and Sekar (2014) add further emphasis to influencing and changing users' behaviour, experience, involvement and efficiency through game thinking and gaming mechanisms. Researchers argue that the emphasis in gamification is on changing players' potential behaviour (Deterding *et al.*, 2011) while games are purely entertainment (Xu *et al.*, 2016; Zichermann & Cunningham, 2011). Nevertheless, both games and gamification need to be fun (Azadegan & Riedel, 2012). Kapp (2012) divides games into structural gamification and content gamification. The former does not contain stories but uses some rewards to motivate players to achieve the goals, while the latter uses stories to enhance engagement. Influencing user behaviour through game design is at the core of gamification.

Gaming motivation

Motivation is critical in gamification. Gaming motivations usually include extrinsic and intrinsic motivation. Intrinsic motivations are often associated with inherent satisfaction and pleasure, to learn or to seek self-actualisation (Brown & Vaughn, 2009; Deci & Ryan, 1985). Extrinsic motivations refer to external rewards, including material rewards (e.g. financial rewards) or psychological rewards (e.g. praise) (Brown, 2007). Rewards are an important gaming mechanism. Internal rewards are included in game rewards; for example, points, badges, leaderboards, unlocking options, virtual presents and so on. External rewards are linked with the real world; for example real presents and opportunities to win cash or a lottery (Deterding *et al.*, 2011). Some advergames, such as Mini Gateway Stockholm, use external rewards to attract players. Zichermann (2011) points out that in order to improve a player's engagement, it is necessary to combine external and internal rewards. Seaborn and Fels (2015) mention that if a game draws heavily on external rewards, players might ignore their gaming experiences and fun; furthermore, once the external rewards stop, players will give up the game. Therefore, intrinsic and extrinsic motivation are both important (Crawford, 2011; Xu *et al.*, 2014). Researchers suggest that players' motivation for playing games mainly focuses on pure entertainment; for example fun, killing time and so on. However, Xu *et al.* (2016) find different motivations for tourist players, they suggest purposive information seeking and socialising are important motivations for tourist players. Happiness, enjoyment, immersion and flow are usually also found in gaming experiences (Ermi & Mäyrä, 2005).

Gamification for tourism marketing

Loyalty cards, stamp books and reward memberships were early forms of gamification (Seaborn & Fels, 2015; Xu *et al.*, 2016). Çeltek (2010) and Afshar *et al.* (2004) suggest that compared with traditional ways of marketing, gamification is a non-intrusive way for marketing and branding, as brands can be placed in games in a less aggressive way (Schneider & Cornwell, 2005). Negruşa *et al.* (2015), meanwhile, suggest gamification can be used as a marketing innovation in the tourism and hospitality sectors.

There are three ways of marketing tourism destinations via gamification (see Table 7.1):

(1) Fully developed tourism games: design a totally new game based on tourism destinations, such as the Austria Adventure Game and Smile Thailand. These types of games are usually developed by Destination management organziations (DMOs) with a particular purpose in order to generate brand awareness in potential customers, to stimulate exploration of the virtual destination and to educate potential tourists

Table 7.1 Different types of gamification in tourism

Type of games	Characteristics	Marketing purpose of the game	Targeting segments	Examples	Developers
Fully developed tourism games	Specifically developed new games for tourism destinations, can be developed by DMOs and other sectors in the tourist industries such as airlines	Brand awareness	Potential tourists	Austria Adventure Game Happy Island	Austria National Tourism Office & Austria Airlines. Irish National Tourism Board
Partially developed tourism games	Using an existing game as a platform, input tourism brand into the existing game	Brand awareness	Potential tourists	Assassin's Creed Wanzhuan Nanjing	UBIsoft Nanjing Tourism Authority
Gamified app	Gamified mobile tour app, usually borrows the idea of location-based games	Enhance on-site tourists' experiences and engagement	On site tourists	RExplore Tripadvisor Whaiwhai travel book	Regensburg Tripadvisor Whaiwhai

DMO: Destination management organization

about the customs of the destination. These games can be used as a supplement to delivering of tourism destination information. Tourists can experience the destination virtually and learn about necessary tourist information through the game; players can also share their experiences with other players, which reduces the uncertainty of travelling.

(2) Partially developed tourism games: using an existing game platform, tourism advertisements are planted into the existing game. For example, 'Wanzhuang Nanjing' uses the existing 'Monopoly' game platform, but inputs the local brand of visitor attractions and restaurants. Some game companies have already collaborated with tourism companies; for example, the historical city of London has been put into the virtual background of the game 'Assassin's Creed'. Some popular tourist attractions, such as Yellow Stone National Park, have been put into the virtual background of the game 'The World of Warcraft'.

(3) Gamified apps: using game thinking and game mechanisms to improve an existing tourism app, to add fun and encourage tourist engagement. For example, Tripadvisor has gamified its app to add some gaming mechanisms – checking, points and sharing have been reported (Sigala, 2015b). Tourists win points when they check in at different destinations.

Academic research on gamification in tourism practices is emerging (Corrêa & Kitano, 2015; Linaza *et al.*, 2013; Negruşa *et al.*, 2014; Shen & Joppe, 2019; Xu *et al.*, 2013). Sigala (2015a) analysed the gamification mechanics adopted by Tripadvisor, proving that gamification can provide a stimulating travel environment for tourists. Bulencea and Egger's (2015) book was the first on gamification in the tourism context. Overall, research on tourism gamification is still limited, and qualitative methods or conceptual discussion are mainly used. Researchers believe gamification can bring different kinds of tourist experiences (Sigala, 2015b), but empirical research is still scarce.

Brand awareness and brand loyalty

Dick and Basu (1994) suggest there are three aspects associated with a brand, which includes cognitive, affective and conative dimensions. Researchers suggest that as a core concept, brand awareness influences consumers' decision-making processes (Campon *et al.*, 2013; Gupta & Kenneth, 1998). Keller (1993) suggests that brand recognition and brand recall are important dimensions of brand awareness. Brand recognition refers to the customer's ability to recognise a brand, while brand recall means consumers' ability to generate the brand from memory. Therefore, in this study, brand recognition and brand recall have been used to measure players' brand awareness. Brand loyalty is another important factor in decision making. Oliver (1999: 34) defines loyalty as,

> a deeply held commitment to re-buy or repatronize a preferred product/service consistently in the future, thereby causing repetitive same-brand or same brand-set purchasing, despite situational influences and marketing efforts having the potential to cause switching behavior.

In the tourism literature, behaviour attention, such as recommending or revisiting the destination (Campon *et al.*, 2013) has often been used as a measurement of loyalty. Therefore, in this study, recommendation of the game, recommendation to visit and intention to visit the destination have been used to measure brand loyalty.

Positive gaming experiences can help the customer concentrate on the brand in the game, recognise the brand and appreciate the brand (Cauberghe & De Pelsmacker, 2010; Xu *et al.*, 2014). Based on the above literature, the following hypotheses were developed:

H1: Playful gaming experiences significantly influence tourist information gained from the game.

H2: Playful gaming experiences significantly influence brand recognition in consumers.

This study explores whether tourism gamification can stimulate tourism buying behaviour, increase brand awareness and buying the brand. Therefore, the following hypotheses were developed:

H3: Playful gaming experiences significantly and positively influence intention to buy.
H4: Brand awareness significantly and positively influences intention to buy.
H5: The tourist information gained from the game significantly and positively influence intention to buy.

Methodology

In order to select a suitable tourism game for the case study, the research team set up three conditions: (1) the game is to be developed by a DMO; (2) the purpose of the game is to promote the destination; and (3) both the internal and external game mechanisms have been used. Eventually, the Austria Adventure Game was selected by the research team.

The Austria Adventure Game was jointly developed by the Austria National Tourism Office and Austria Airlines and released officially in 2015. Through a 3D game background, the game illustrates the natural beauty the picturesque scenery of Austria so players can experience the destination virtually. In addition, through the face recognition function, the game enhances players' interest and sense of novelty. The Austria Adventure Game is a challenge game, because it uses both internal gaming mechanisms (e.g. self-challenge, leaderboards) and external mechanisms, such as rewards; for example, the winners of the game will have the chance to be randomly selected by the Austria National Tourism Office to enjoy a free trip to Austria, as long as the game player provides screenshots of the game and posts it on social media networking sites to share with friends (Austria National Tourist Office, 2016).

This study uses a quantitative method through the use of questionnaires to explore the relationship between gaming experiences and tourism marketing, and discusses whether gamification marketing can increase tourist players' brand awareness and buying potential. Players were given the Austria Adventure Game to play first; after playing the game, they were asked to fill in the questionnaire. The questions were divided into four parts: (1) player motivation, including internal and external motivation (Boyle *et al.*, 2012; Deterding *et al.*, 2011; Ryan *et al.*, 2006); (2) player experiences, including general gaming experience and immersion (Ermi & Mäyrä, 2005) (e.g. I feel totally immersed in the game, I feel myself part of the virtual environment); (3) tourism brand awareness, including brand awareness and brand identity (Keller, 1993); and (4) tourist intention to buy, including items such as 'because of this game, I would go to Austria'. Other details, such as 'through this game, I would like to know more about tourism information in Australia', 'I have an increased interest in going to Austria; this game has changed my existing tourism plans to visit Austria', 'I would recommend this game to others' and 'I agree game-based marketing is a good idea' were also asked to gain a more general overview. A 1 to 5 Likert scale was used to grade the answers, with 5 being totally agree and 1 totally disagree.

A pilot test was conducted based on a convenience sample of 15 game players to clarify the wording; a few sentences were then reworded based on the feedback. The final version of the questionnaire was conducted in Southeast University, Nanjing, China. Southeast University is a comprehensive university with 18,000 enrolled undergraduate, postgraduate and PhD students. A convenient sampling method was used at the Student Centre on campus, where volunteer participants were recruited during a week in June 2016. Most participants were students (85.3%), with 14.7% being university staff members. Williams *et al.* (2008) suggests game players are usually aged under 30, and university students are usually technology driven and willing to accept new things therefore, student samples are frequently used in gaming-related studies (Arif *et al.*, 2016; Griffiths *et al.*, 2004). Each participant was provided with an iPad or iPhone that had been pre-loaded with the Austria Adventure Game, and each participant was asked to play the game until they decided to quit freely. After the playing the game, each participant was asked to fill in the questionnaire based on their experiences of the game. In total, 155 people volunteered in one week, and in the end 131 valid questionnaires were collected. Data were evaluated using SPSS 22.

Results

Most participants were university students (86.3%) and were aged between 18 and 25 years old. Males accounted for 41.2% and females 58.8% of the participants. Most participants held an undergraduate degree or were studying on an undergraduate programme (91.6%); 87.8% participants earned below 3000 RMB per month. The average time spent playing the Austria Adventure Game was 12.95 minutes, with a median of 10 minutes; the longest time spent playing the game was 30 minutes and lowest time spent was 4 minutes. A validity and reliability test suggested the Cronbach alpha was 0.899, suggesting a reasonable result.

Motivation to play the game

Mean values suggest the top three motivations for playing the game included both internal and external motivations, and all had a clear association with travelling. The top motivation was looking for travel information about Austria (mean = 3.14), followed by enjoying the virtual travelling experience (mean = 2.92) and the third motivation was looking for rewards to go to Austria (mean = 2.74). These top three gaming motivations have a clear association with travel and tourism, which differs from traditional gaming motivations such as killing time and challenging oneself (Chou & Tsai, 2007; Lucas & Sherry, 2004). This result is similar to that of Xu *et al.* (2016) who found that the motivations of tourism game players were mainly curiosity and seeking tourism information, indicating that tourism games have their own features and they can be attractive to potential tourists. It is obvious from the results that both internal and external motivations are important, supporting the findings of Deterding *et al.* (2011) and Zichermann (2011).

Descriptive results of the mean values suggest respondents showed a great interest in the information gained when playing tourism games. For example, 'through this game, I have increased interest to go to Austria' (mean = 2.92, median = 3, df = 1.130); 'through this game, I have learned more tourism related information in Austria' (mean = 3.02, median = 3, df = 1.034); and 'through this game, I have gained enough tourism related information' (mean = 3.04, median = 3, df = 1.011). Players are interested in the tourism information presented in the game and they showed a high agreement with game-based marketing (mean = 3.97).

Tourism experiences and brand

First, factor analysis was used to reduce dimensions. There were 11 items on tourism experiences and factor analysis result showed three factors (KMO = 0.779, factor loading 64.248%): factor 1 was named 'general gaming experiences', which included 'I am attracted by the game design', 'the game is interesting', 'the game is fun', 'I feel happy in the game' and 'I enjoy my experience in the game'. Factor 2 was named tourism information, which includes 'purposively looking for tourism information in the game' and 'the game increased my knowledge/information of Austria'. Factor 3 was named virtual experience, which included 'I feel myself as part of the virtual environment' and 'I feel totally immersed in the game'. One item had double loading ('I remember the virtual environment in the game'); therefore, this item was deleted. A second run of factor analysis suggested all results had been improved (see Table 7.2).

There are seven items regarding tourism branding and factor analysis results only show one factor, that is brand awareness and loyalty (KMO = 0.829, factor loading 55.353%), which includes: 'I can clearly remember the brand in the game', 'I have good feelings towards the brand', 'I have a deep impression of the brand', 'I can recognise the tourism brand', 'I can use/visit the brand when I travel to Austria', 'I would recommend the brand to others' and 'I would try to use the brand several times' (see Table 7.2 for details).

Table 7.2 Factor loading of latent variables

	Item	Factor loading	KMO
General gaming experiences F1	I am attracted by the game design	0.738	0.779
	The game is interesting	0.745	
	The game is fun	0.814	
	I feel happy in the game	0.854	
	I enjoy my experience in the game	0.829	
Tourism information F2	Purposively looking for tourism information in the game	0.684	
	The game increased my knowledge/information of Austria	0.775	
Virtual experience F3	I feel myself as part of the virtual environment	0.847	
	I feel totally immersed in the game	0.648	
Brand F4	I can clearly remember the brand in the game	0.713	0.829
	I have good feelings towards the brand	0.812	
	I have a deep impression of the brand	0.751	
	I can recognise the tourism brand	0.791	
	I can use/visit the brand when I travel to Austria	0.762	
	I would recommend the brand to others	0.699	
	I would try to use the brand several times	0.669	

KMO: Kasier-Meyer-Olkin test

Gaming experiences, brand awareness and travel intention

Next, a linear regression model was used to test the relationship of each factor (independent variable) and the item 'I would go to Austria because of this game' (dependent variable). Five models were established (see Table 7.3).

Table 7.3 suggests that gaming experiences have a significant and positive relationship with the intention to buy (in this case travelling to Austria), meaning players' positive gaming experience could significantly improve the intention to travel to Austria. Therefore, H3 was supported.

Brand awareness and brand loyalty significantly and positively influence intention to buy, namely, players who have a high brand awareness and brand loyalty would have a high intention to travel to Austria. Therefore H4 was supported.

The tourist information gained from the game significantly and positively influenced the intention to travel to Austria. Tourists who were satisfied with the tourist information gained in the game would have a high intention to visit Austria. Therefore H5 was supported.

Table 7.3 The linear regression model of gaming experience and tourism brand

Independent variable	Model 1	Model 2	Model 3	Model 4	Model 5
Gaming experiences F1	$R^2 = 0.081$, F = 12.508, Sig = 0.001				
Tourism Information F2		$R^2 = 0.06$, F = 9.908, Sig = 0.002			
Virtual experience F3			$R^2 = 0.01$, F = 2.382, Sig = 0.125		
Brand F4				$R^2 = 0.122$, F = 19.125, Sig = 0.000	
F1, F2, F3, F4 + age + gender + play time					$R^2 = 0.253$, F = 7.274, Sig = 0.000

Note: Dependent variable is 'I would go to Austria because of this game'

However, the gaming experience does not have a positive correlation with brand awareness and brand loyalty. ($R^2 = 0.026$, F = 4.422, Sig = 1.000). Therefore, H2 was rejected. Gaming experience does not have a positive correlation with tourism information either, therefore, H1 was rejected as well. Table 7.4 summarises the results of the hypotheses testing.

Table 7.4 Results of hypotheses testing

Hypotheses	Test	Test results	Hypotheses results
H1: Playful gaming experiences significantly influence tourist information gained from the game	Liner regression	$R^2 = 0.034$, F = 3.651, Sig = 1.023	Rejected
H2: Playful gaming experiences significantly influence the brand recognition of consumers	Liner regression	$R^2 = 0.026$, F = 4.422, Sig = 1.000	Rejected
H3: Playful gaming experiences significantly and positively influence intention to buy	Liner regression	$R^2 = 0.081$, F = 12.508, Sig = 0.001	Supported
H4: Brand awareness significantly and positively influence intention to buy	Liner regression	$R^2 = 0.122$, F = 19.125, Sig = 0.000	Supported
H5: The tourist information gained from the game significantly and positively influence intention to buy	Liner regression	$R^2 = 0.06$, F = 9.908, Sig = 0.002	Supported

Social demographic factors and travel intention to Austria

In order to test the influence of each factor of 'intention to buy (in this case travelling to Austria)', factors such as age and gender were also put into the regression model together with variable 'playing time' after being converted into dummy variables. A linear regression model of this complex set of variables (independent variable) and the willingness to purchase variable ('I would go to Austria because of this game') suggest $R^2 = 0.253$, $F = 7.274$, $Sig = 0.000$, while the variables 'gaming experience' and 'brand' would significantly influence intention to buy 'I would go to Austria because of this game' (Table 7.5). However, social demographic variables did not have a significant influence on travel intention, while brand awareness and game experiences have proven to have significant influence on travel intention.

Brand awareness and brand recognition significantly and positively influence intention to buy, namely, the tourism-related brand awareness and brand recognition established in the game influences the intention to buy in tourism. While tourism information in the game can significantly and positively influence brand awareness and brand recognition, tourism information established in the game can also contribute to a positive brand awareness and brand recognition. The virtual experiences in the game do not directly contribute to intention to buy, but virtual experiences significantly influence brand awareness and brand recognition; namely, if the player immerses themselves in virtual Austria, they have a high possibility of building a positive brand awareness and brand

Table 7.5 Regression model of intention to buy and latent variables*

Model	Unstandardised coefficients		Standardised coefficients	*t*	Sig
	B	Standard error	Beta		
Constant	3.043	0.356		8.547	0.000
Virtual experiences	−0.0058	0.106	−0.048	−0.551	0.583
Game experiences	−0.0488	0.098	−0.399	−4.967	0.000
Tourism information	0.130	0.104	0.106	1.250	0.214
Brand	0.464	0.114	0.380	4.082	0.000
Gender	−0.074	0.190	−0.030	−0.389	0.698
Age	0.276	0.345	0.063	0.801	0.425
Minutes spent on the Austria Adventure Game	−0.021	0.013	−0.126	−1.560	0.121

* Dependent variable: 'Because of this game, I would go to Austria'. Independent variables: virtual experiences, game experiences, tourism information, brand, gender, age and minutes spent on Austria Adventure Game

recognition in the game. The results also suggests that gaming experiences influence the intention to buy, meaning a positive experience in the game would increase intention to buy (in this case, travel to Austria). In this study, positive gaming experiences do not have a relationship with brand awareness and brand recognition, which is different from Cauberghe and De Pelsmacker (2010) who propose gaming experiences influence brand recognition.

Travel information from the game and travel intention to Austria

A linear regression model of the item 'I would go to Austria because of this game' and items on travel decision-making were established. For example, 'I would like to know about Austria through this game' (R^2 = 0.000, F = 0.058, sig = 0.306); 'This game can influence and change my travel plans to Austria' (R^2 = 0.262, F = 47.118, sig = 0.000); 'Through this game, I have increased interest to travel to Austria' (R^2 = –0.007, F = 0.064, sig = 0.801); 'I have learned more travel information on Austria through this game' (R^2 = 0.014, F = 2.817, sig = 0.096); and 'I have learned enough information about Austria' (R^2 = 0.086, F = 13.257, sig = 0.000). Results suggest that the item 'I have learned enough information about Austria' significantly and positively influenced 'I would go to Austria because of this game'; namely, if the tourist is satisfied with the tourism information, they would have a high intention to visit the destination. Also, the item 'This game can influence and change my travel plans' significantly and positively influences 'I would go to Austria because of this game' (R^2 = 0.262, F = 47.118, sig = 0.000), meaning that people who agree tourism information gained in the game can influence their travel plan would have a high intention to travel to the destination.

These results suggest the importance of providing travel information in the game, as this will have a great influence on travel intentions.

Conclusions

Gamification, as a new type of marketing tool, has been adopted by an increasing number of businesses. Tourism destinations have also incorporated this type of marketing into their businesses in order to build brand awareness and enhance tourism experiences and involvement. However, there is limited empirical study on how gamification influences tourism destination marketing. Based on a case study on the Austria Adventure Game, this exploratory research used a questionnaire and discussed how tourism gamification can enhance tourism brand building and how gamification can influence players' intentions to visit a destination.

The results suggest tourism brand awareness and brand recognition through the game can contribute to intention to travel to the destination. Tourism information gained in the game can contribute to a positive brand awareness and brand recognition. Although virtual experiences in the game did not have a significant influence on the intention to visit the destination, they contributed to brand awareness and brand recognition; namely, players who have had a positive virtual experience in the game demonstrated a positive brand awareness and brand recognition. Therefore, this study shows that gamification as a new marketing tool can have some positive effect on tourists' decision-making and provides empirical evidence for other researchers (Sigala, 2015b; Xu *et al.*, 2014).

The results also suggest that tourist players' motivations are different from traditional game players, with a significant number of participants purposively seeking travel information, experiencing the destination virtually and looking for rewards to go to the destination. These results confirm Xu *et al.*'s (2016) study. These motivations can be linked to the push factors of travelling.

Gamification can be used by tourism destinations as a new marketing tool to communicate tourism information. Through deliberately developed tourism games, destinations can establish relevant tourism brands in the destination and stimulate potential visits through the tourism information in the game.

References

Afshar, R., Jones, C. and Banerjee, D. (2004) *Advergaming Developer's Guide: Using Macromedia Flash MX 2004 and Director MX*. Hingham, MA: Charles River Media.

Arif, I., Aslam, W. and Ali, M. (2014) Students' dependence on smart-phones and its effect on purchase behavior. *South Asian Journal of Global Business Research* 5 (2), 285–302.

Austria National Tourist Office (2016) Austria snapshot adventure. Available at: http://www.austria.info/in/austria-snapshot-adventure (accessed 17 November 2017) .

Azadegan, A. and Riedel, J.C.K.H. (2012) Serious games integration in companies: A research and application framework. Proceedings of the 12th International Conference on Advanced Learning Technologies. IEEE, 04–06 July, Rome, Italy, pp. 485–487.

Bogost, I. (2007) *Persuasive Games: The Expressive Power of Videogames*. Cambridge, MA: MIT Press.

Boyle, E.A., Connelly, T.M., Hainey, T. and Boyle, J.M. (2012) Engagement in digital entertainment games. *Computers in Human Behavior* 28, 771–780.

Brown, S.L. and Vaughn, C.C. (2009) *Play: How it Shapes the Brain, Opens the Imagination and Invigorates the Soul*. New York, NY: A Very Publishing Group.

Bulencea, P. and Egger, R. (2015) *Gamification in Tourism: Designing Memorable Experiences*. Norderstedt: Books on Demand.

Campon, A.M., Alves, H. and Hernandez, J.M. (2013) Loyalty measurement in tourism: A theoretical reflection. In A. Matitas, P. Nijkamp and M. Sarmento (eds) *Quantitative Methods in Tourism Economics* (pp. 13–40). New Mexico, US: Springer.

Cauberghe, V. and De Pelsmacker, P. (2010) ADVERGAMES: The impact of brand prominence and game repetition on brand responses. *Journal of Advertising* 9 (1), 5–18.

Çeltek, E. (2010) Mobile advergames in tourism marketing. *Journal of Vacation Marketing* 16 (4), 267–281.

Chaffery, D. and Ellis-Chadwick, F. (2012) *Digital Marketing, Strategy, Implementation and Practice* (5th edn). London: Pearson.

Chou, C. and Tsai, M.J. (2007) Gender differences in Taiwan high school students' computer game playing. *Computers in Human Behavior* 23, 812–824.

Corrêa, C. and Kitano, C. (2015) Gamification in tourism: Analysis of Brazil Quest Game. *e-Review of Tourism Research* 6, 1–5.

Crawford, C. (2011) *The Art of Computer Game Design* (2nd edn). Seattle: Amazon Digital Services, Inc.

Deci, E.L. and Ryan, R.M. (1985) *Intrinsic Motivation and Self-determination in Human Behavior*. New York: Plenum Publishing.

Deterding, S., Dixon, D., Khaled, R. and Nacke, L.E. (2011) From game design elements to gamefulness: Defining 'gamification'. *MindTrek Proceedings*, Tampere, Finland: ACM Press.

Dick, A. and Basu, K. (1994) Customer loyalty: Toward an integrated conceptual framework. *Journal of Academy Market Science* 22 (2), 99–133.

Epstein, Z. (2013) Enterprise gamification for employee engagement. Available at: https://doksi.net/en/get.php?lid=29219 (accessed 23 December 2020).

Ermi, L. and Mäyrä, F. (2005) Player-centred game design: Experiences in using scenario study to inform mobile game design. *Game Studies* 5 (1). http://www.gamestudies.org/0501/ermi_mayra/ (accessed 23 December 2020).

Griffiths, M.D., Davies, M.N.O. and Chappell, D. (2004) Demographic factors and playing variables in online computer gaming. *Cyber Psychology & Behavior* 7 (4), 479–487.

Gupta, P.B. and Kenneth, R.L. (1998) Product placement in movies: The effect of prominence and mode on audience recall. *Journal of Current Issues and Research in Advertising* 20 (1), 47–59.

Jennings, J.E., Devereaux Jennings, P. and Greenwood, R. (2009) Novelty and new firm performance: The case of employment systems in knowledge-intensive service organizations. *Journal of Business Venturing* 24, 338–359.

Jobber, D. (2009) *Principles and Practices of Marketing* (6th edn). Berkshire: McGraw-Hill Higher Education.

Kapp, K.M. (2012) *The Gamification of Learning and Instruction: Game Based Methods and Strategies for Training and Education*. San Francisco, CA: Pfeiffer.

Keller, K.L. (1993) Conceptualizing, measuring, and managing customer-based brand equity. *Journal of Marketing* 57 (January), 1–22.

Linaza, M.T., Gutierrez, A. and García, A. (2013) Pervasive augmented reality games to experience tourism destinations. In Xiang, Z. and Tussyadiah, I. (eds) *Information and Communication Technologies in Tourism 2014* (pp. 497–509). Switzerland: Springer International Publishing.

Lucas, K. and Sherry, J. (2004) Sex differences in video game play: A communication-based explanation. *Communication Research* 31 (5), 499–523.

Middleton, V.T.C., Fyall, A. and Morgan, M. (2009) *Marketing in Travel and Tourism* (4th edn). Oxford: Butterworth-Heinemann.

Negruşa, L.A., Toader, V., Sofică, A., Tutunea, F.M. and Rus, V.R. (2015) Exploring gamification techniques and applications for sustainable tourism. *Sustainability* 7 (8), 11160–11189.

Oliver, R. (1999) Whence Consumer Loyalty. *Journal of Marketing* 63 (4), 33–44.

Pine, B.J. and Gilmore, J.H. (1999) *The Experience Economy: Work is Theatre and Every Business is a Stage*. Boston, MA: HBS Press.

Ryan, R.M., Rigby, C.S. and Przybylski, A. (2006) The motivational pull of video games: A self-determination theory approach. *Motivation and Emotion* 30, 347–363.

Schneider, L-P. and Cornwell, B.B. (2005) Cashing in crashes via brand placement in computer games. *International Journal of Advertising* 24 (3), 321–343.

Seaborn, K. and Fels, D.I. (2015) Gamification in theory and action: A survey. *International Journal of Human-Computer Studies* 74, 14–31.

Sigala, M. (2015a) The application and impact of gamification funware on trip planning and experiences: The case of TripAdvisor's funware. *Electron Markets* 25, 189–209.

Sigala, M. (2015b) Gamification for crowdsourcing marketing practices: Applications and benefits in tourism. In F.J. Garrigos-Simon, I. Gil-Pechuán, S. Estelles-Miguel *et al.* (eds) *Advances in Crowd Souring* (pp. 129–145). Cham: Springer.

Stadler, D. and Bilgram, V. (2016) Gamification: Best practices in research and tourism. In R. Egger, I. Gula, D. Walcher *et al.* (eds). *Open tourism, tourism on the verge.* Springer-Verlag: Berlin Heidberg. doi: 10/1007/978-3-642-54089-9_28.

Uskov, A. and Sekar, B. (2014) Serious games, gamification and game engines to support framework activities in engineering: Case studies, analysis, classifications and outcomes. Proceedings of the 2014 IEEE International Conference Electro/Information Technology (EIT), 5–7 June 2014, Milwaukee, WI, USA, pp. 618–623.

Waltz, S.P. and Ballagas, R. (2007) Pervasive persuasive: A rhetorical design approach to a location-based spell-casting game for tourists. Proceedings of Situated Play. DiGRA 2007: The 3rd International Digital Games Research Conference, 24–28 September 2007, Tokyo, 489–497.

Williams, D., Yee, N. and Caplan, S.E. (2008) Who plays, how much, and why? Debunking the stereotypical gamer profile. *Journal of Computer-Mediated Communication* 13 (4), 993–1018.

Xu, F., Tian, F., Buhalis, D. and Weber, J. (2013) Marketing tourism via electronic games: Understanding the motivation of tourist players. 5th International Conference on Games and Virtual Worlds for Serious Applications, Bournemouth, 11–13 September, UK.

Xu, F., Weber, J. and Buhalis, D. (2014) The gamification of tourism. In Z. Xiang and I. Tussyadiah (eds) *Information and Communication Technologies in Tourism 2014* (pp. 525–537). Wien: Springer. Proceedings of the International Conference, 21–24 January, Dublin.

Xu, F., Tian, F., Buhalis, D., Weber, J. and Zhang, H. (2016) Tourists as mobile gamers: Gamification for tourism marketing. *Journal of Travel and Tourism Marketing* 33 (8), 1124–1142.

Xu, F., Buhalis, D. and Weber, J. (2017) Serious games and the gamification of tourism. *Tourism Management* 60, 244–256.

Zichermann, G. (2011) Intrinsic and extrinsic motivation in gamification. Gamification Co. Available at: https://www.gamification.co/2011/10/27/intrinsic-and-extrinsic-motivation-in-gamification/ (accessed 23 December 2020).

Zichermann, G. and Linder, J. (2010) *Game Based Marketing.* Hoboken, NJ: John Wiley & Sons Inc.

Zichermann, G. and Cunningham, C. (2011) *Gamification by Design: Implementing Game Mechanics in Web and Mobile Apps.* Newton, MA: O'Reilly Media.

8 Gamification and Geocaching for Tourism Destinations: Marketing Madeira, Portugal

Luiz Pinto Machado

Introduction

The success of a tourist destination depends on the tourist experiences that take place there. Many destinations seek to present genuine experiences that involve the tourist in an oriented action, combining discovery and pleasure dimensions associated with physical and mental well-being. Destinations can allow for challenges that combine adventure with good climate, landscape and hospitality. They can offer unforgettable experiences to the 'geocachers', a niche of 'gamification', which is a growing trend throughout the world and has become a focus of attention in an increasing number of fields, including tourism businesses. During recent years, the number of caches in Madeira Island has grown. Contrary to the mainland, geocachers are mostly tourists that do not have enough time to search for 'caches' in a relaxed way. As geocachers are tourists, they plan round trips and look not for only one cache but a string of caches in which they can vary the order individually. Therefore Destination Management Organizations (DMOs) should include in their official web pages some specific information that helps with finding geocache places, alongside photographs of geocaching experiences, and develop apps and games in an accessible format. This could be a way to attract not just geocachers, but tourists in general, by showing unique and unfamiliar places and proposing new challenges. This chapter seeks to identify how the development of the 'gamification' concept and specifically 'geocaching' can contribute to attracting more tourists and to reinforcing the destination image of Madeira. The chapter further describes the limitations of gamification and suggests future research on the topic. Policy implications are also discussed.

Framework

> Like making music, telling stories, and creating images, playing games is part of what it means to be human. Games are perhaps the first designed interactive systems our species invented. The rise of computers has paralleled the resurgence of games in our culture. This is no accident. Games like Chess, Go, and Parchisi are much like digital computers, machines for creating and storing numerical states.... In this sense, computers didn't create games; games created computers. (Zimmerman, 2013)

It is generally accepted that the world, industry and human beings depend on technology, and tourism is no exception. Future tourists will be part of the Millennial generation; they were born to use the technology and depend on it. They are a generation that cannot live without a mobile phone, that has the world in the 'palm of the hand' and that works in a network. They will travel more, spend more money and have great influence on other tourists in the destination choice process.

Tourism, as a service-intense industry, has gone for many decades hand-in-hand with technology and embraced the potential inherent in its latest developments (Buhalis & Law, 2008). However, what has changed significantly is that technology has not only become an integral part of tourism, but it has also revolutionised the way travel is planned (Buhalis, 2003), how business is conducted (Buhalis & Licata, 2002) and how tourism services and experiences are created and consumed (Stamboulis & Skayannis, 2003). This has opened up new opportunities, challenges and potential in the field (Gretzel *et al.*, 2006).

Those in the tourism and hospitality industries at tourist destinations should understand the behaviour of this new generation as an opportunity to make destinations and tourism companies more competitive by creating memorable experiences that are associated with technology. These creations serve to reinforce the motivation to choose a destination and to increase tourist satisfaction. Consequently, they contribute to the improvement of the reputation and image of the destination. Films, poetry and literature have had a strong influence on past generations in the formation of their identity. For new generations, games play a highly important role in their perception and behaviour. Games are dynamic systems based on interactions that help human beings learn from the first years of life. Zimmerman (2013), a game designer, wrote a provocative essay entitled 'Manifesto for a Ludic Century', in which he declares,

> In the Ludic Century, we cannot have a passive relationship to the systems that we inhabit. We must learn to be designers, to recognize how and why systems are constructed, and to try to make them better. These are not just aspects of the Ludic Century, they are also elements of *gaming literacy*. Literacy is about creating and understanding meaning, which allows people to write (create) and read (understand). How dynamic interactive systems create beauty and meaning – is one of the delightful and daunting challenges we face in this dawning Ludic Century.

The concept of gamification is defined as the use of game design elements and game thinking in a non-gaming context (Deterding *et al.*, 2011), ranging from increasing brand awareness to encouraging consumer engagement. Inside the huge scope of this concept, the activity of geocaching fits perfectly. Applications are used in a range of contexts, from education and training to health, self-management, innovation, employee engagement, heritage, crowdsourcing, civic engagement and marketing (Seaborn & Fels, 2015). There has been a high growth of gamification solutions among small and medium enterprises in industry verticals owing to an increasing need for gamification solutions and applications in enterprise and consumer brands. The gamification market size is projected to grow from USD 9.1 billion in 2020 to USD 30.7 billion by 2025, at a Compound Annual Growth Rate (CAGR) of 27.4% during the forecast period. The major factors driving the growth of the gamification market include rewards and recognition to employees over performance to boost employee engagement, provision of lucrative offers to the customers and consumers, and gamification yielding higher Return on Investment (RoI) (Markets & Markets, 2020). The experience economy hit the zeitgeist as a key instrument to yield differentiation, added value and competitive advantage. In practice, the principal idea for businesses was to no longer compete in terms of price but in terms of the distinctive value of the experience provided. For the years to follow, the experience economy provided an unprecedented guide for strategic staging, managing and delivering experiences to consumers in a variety of contexts and industries (Neuhofer & Buhalis, 2013). Particularly fostered by the adoption of emerging technologies, such as interactive games, online spaces and virtual reality, it was possible to meet the demand and create ever-more immersive consumer experiences (Pine & Gilmore, 1998).

This rapid technological development has led to massive mobility in terms of the physical movement of products, services and people and at the same time encouraged the mobility and ubiquity of technological artifacts themselves (Gretzel & Jamal, 2009). Mobility has been identified as one of the four big trends next to globalisation, communication and virtuality, as identified by Egger and Buhalis (2008).

This study aims to analyse the recent attempts to rejuvenate a classic mature destination (Madeira Island) based on the development of 'alternative' concepts. This is by assessing the real prospects of developing new and genuine experiences that involve the tourist in an oriented action, combining discovery and pleasure dimensions associated with physical and mental well-being. There is an obvious interest, shared by many classic destinations, in recovering lost ground in historical markets.

The motivation for the present research is to analyse the 'Ludic Century' phenomenon in general and particularly gamification. There is a tradition of analysing tourism in classic destinations, but gamification has received limited research attention.

Considering that the 'nature' element, complemented by culture and gastronomy, is the model supported by the official strategic studies for the island's tourism development, at least until 2021 (Secretaria Regional da Economia Turismo e Cultura, 2016), allied with the fact that the ideal age group is signalled to be between 45 and 55 years of age, the aim of this chapter is to show through the study carried out in the east side of the island (and its results), that geocaching fits perfectly into the island's tourism development strategy for the coming years. Moreover, classic mature destinations should increase the number of tourists, at any time of the year, based on the development of 'alternative' concepts by assessing the real prospects of developing new and genuine experiences that involve the tourist in an oriented action.

As geocaching continues to evolve as a very dynamic area of research enabling innovation again in this context, this chapter can also contribute to the reinvention of classic destinations.

The Tourism Sector as Play and Game

The tourism industry is one of the human social areas that has most benefited from technological progress and much of its progress over the years is due to this connection. However, there is still a lot to do to be part of the Ludic Century. The application of game design elements in tourism is already present, but it might not have been recognised or been named as the concept of gamification (Xu *et al.*, 2013).

Game design in tourism is something that researchers should study, to understand both the supply and demand sides, and to enable innovation and reinvention of classic destinations.

Tourism is no longer stationary, but experiential, creating an ever-closer bond among its participants. Visitors no longer want to 'look out of the window of a bus', but be able to touch, feel and sample the environment around them. They want to be able to interact with the locals, learn about their culture first-hand, seek greater understanding of the communities that ultimately make the attraction come alive, trough co-creating experiences.

The same principles extend to other tourist attractions. As an example, Madeira Island, a popular year-round tourist destination since the 1890s, is at the forefront of providing such excitement. It is not only one of the best European spots for hiking, with its unique 'levada hike system', but it also offers great cultural heritage attractions, that combined with other adventure items, can improve their experiential elements.

In tourism marketing, the concept of the experience economy has long provided a valuable vehicle to design, stage and deliver experiences to consumers and gain competitive advantage. The proliferation of recent information and communication technologies (ICTs) has had a great influence on society and people's everyday lives and has implied new challenges for the creation of experiences (Neuhofer & Buhalis, 2014). With the proliferation of ICTs, the potential for experiences to be co-created has 'exploded on an unprecedented scale everywhere in the value creation system' (Ramaswamy, 2009). Neuhofer and Buhalis (2013) argue that through ICTs, co-creation is no longer restricted to companies and consumers (B2C) but is enabled among consumers and social networks (C2C) on all levels. They say that one of the possible applications of ICTs it to allow tourists to experience the physical tourist environment and stay connected in the online space at the same time, suggesting exploring the potential of technology as a key tool to facilitate more opportunities for social co-creation experiences for the tourist consumer. Thereby, it is of particular importance to support consumers to co-create experiences outside the company domain with each other (C2C). This means that first and foremost the technological requirements need to be fulfilled that allow the tourist to be connected, for instance, through wireless access in hotel rooms, at airports and wireless destinations.

Destinations with unspoilt nature, which allow challenges that combine adventure with good climate, landscape and the hospitality of its people, can offer unforgettable experiences to geocachers (geocaching practitioners), a niche of gamification that is growing throughout the world and has become a focus of attention in an increasing number of fields, including tourism business.

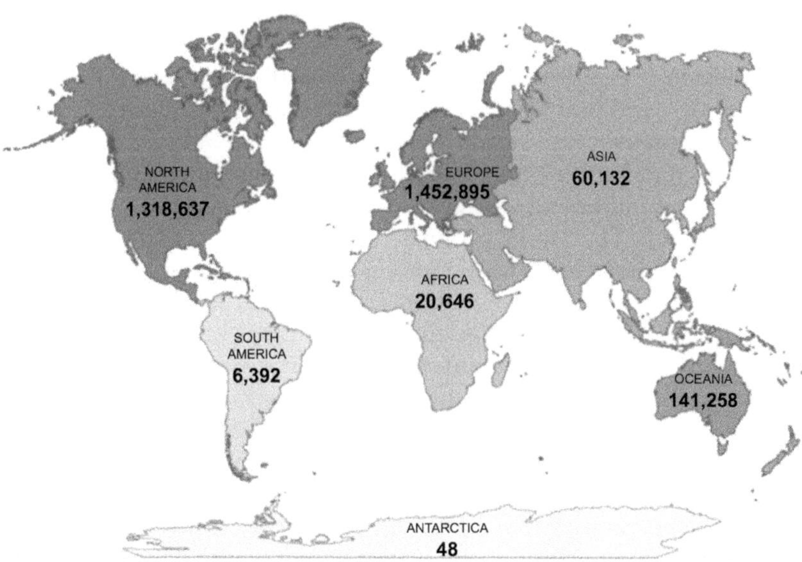

Figure 8.1 Number of caches per continent (April 2018)
Source: Geocaching.com

The number of caches in Madeira Island have been growing since 2015 (Figure 8.1). Contrary to mainland Portugal, geocachers are mostly tourists that do not have enough time to search for caches in a relaxed way. A 'did not find' is always a disappointment. That is why sometimes a spoiler photo, too much rather than too little, is shown in the cache description. As geocachers are tourists, they plan round trips, and look for a string of caches in which they can vary the order individually.

DMOs should include in their official web pages some specific information such as setting-up a web page that helps tourists to find geocache locations, placed alongside a few photo experiences. They should develop some applications and games in accessible formats that are compatible with popular programs such as Google Earth, Garmin MapSource, Kompass and so on.

This could be a way to attract not just the geocachers, but also tourists in general, by showing unique and unfamiliar places and propose new challenges. Moreover, the corporate incentives market is responding very well to this new concept, since it allows, in addition to the discovery of unspoilt nature and heritage, a challenge that implies physical and intellectual competences that can both be used as an element of leisure and team evaluation and training in an outdoor environment.

Despite being a recent theme in tourism, applications are beginning developed. These applications, based on interactive platforms, are highly intuitive and widely used, particularly by large companies in the tourism sector. They often use them as a tourist loyalty technique (e.g. frequent flyer programmes that accumulate points). Gamification has already been used successfully in marketing, business, health and education (Xu *et al.*, 2013). However, the tourism experience literature does not cover the essential dimensions of intrinsically interactive motivation, social play, challenge, fantasy and fun, which is provoked by using game design elements and game thinking.

Buhalis and Foerste's (2015) study connects the different concepts of context-based marketing, social media and personalisation, as well as mobile devices. The model proposes social context mobile (SoCoMo) marketing as a new framework that enables marketers to increase value for all stakeholders at the destination. Contextual information is increasingly relevant, as big data collected by a wide range of sensors in a smart destination provide real-time information that can influence the tourist experience. SoCoMo marketing introduces a new paradigm for travel and tourism. It enables tourism organisations and destinations to revolutionise their offering and to co-create products and services dynamically with their consumers. The proposed SoCoMo conceptual model explores the emerging opportunities and challenges for all stakeholders.

Geocaching, as a tourism and leisure practice, which is growing particularly in Europe and North America, expresses how the use of new positioning techniques has changed personal territorialities, leading its practitioners to contact places and landscapes which so far revealed little centrality and attractiveness (Fernandes, 2012).

The most challenging and unpredictable geocaches are usually found in the woods. They may involve hikes of several miles, in some cases, and finding even large containers can be challenging in dense underbrush or poor light. Furthermore, GPS devices do not like thick woods. Signals may suddenly drop or become unreliable and signals bouncing off rock walls can suddenly render them almost useless (Gillin & Gillin, 2010).

Once players find the cache, they sign their player name into the logbook along with the date of the find. The player can also leave items such as travel bugs that move from cache to cache and accumulate distance along with accounts of the item's travels by the gamers who pass the travel bug along. There are also usually trinkets left by gamers, who can trade one item for another. The last step of the player's experience with a cache is to log the visit online at Geocaching.com. Here, players describe the events of the hunt (often discussing the roles they had to play to not be spotted by non-gamers), the condition of the cache, and the objects left or traded. This multi-step process makes geocaching a game that utilises the correlation between the material landscape and digital space, and depends on the collaboration between these spaces for a sense of embodiment (Farman, 2012).

Digitalisation of Tourism and Gamification

The technological revolution that is centred around information and communication technologies imposes a new rhythm on society, leading to new ways of producing, communicating and managing. With this technological paradigm, people will look for new spaces of leisure and entertainment. But companies in the tourism industry will also use ICTs to market their products and services. At present, there is a growing use of ICTs by operators and consumers, as well as the internet, for the distribution, marketing and acquisition of travel and tourism. ICTs are the most critical areas for tourism success in the future and for how tourist destinations will be promoted (Machado & Almeida, 2010).

The rise of e-tourism (Buhalis & Law, 2008) and new technologies' integration, such as virtual communities (Wang *et al.*, 2001), collaborative websites (O'Connor, 2008) like Tripadvisor or social media (Barabel, 2010) and blogs (Lin & Huang, 2006), have largely contributed to the growth of the tourist sector where the consumer is always looking for more information (Doolin *et al.*, 2002). Technological advances are causing fundamental disruptions in tourism by empowering (traditional but also new) tourism actors to form new markets, offerings, management practices and competitive strategies (Sigala, 2018a). Thus, technologies are transforming tourism management and marketing from a static and utilitarian sense (whereby managers and tourists use technologies as tools) to a transformative conceptualisation whereby tourism markets and actors (tourism providers, stakeholders, intermediaries and tourists) both shape and are shaped by technology (Sigala, 2018b).

Tourism marketing also benefit from virtual reality (VR) and augmented reality (AR), which, when well used, constitutes an opportunity to positively change tourist behaviour in the decision-making process of choosing a destination.

Tussyadiah *et al.* (2017) support the idea that the rapid development of VR technology offers opportunities for a widespread consumption of VR tourism content. The study also presents challenges to better understand the effectiveness of VR experiences in inducing more favourable attitudes towards tourism destinations and shaping visitation intention. That is, a higher sense of presence during VR experiences leads to stronger interest and liking towards the destinations. Recently, much attention has been directed towards AR interfaces. AR browsers deliver (geo)spatial and attribute information about physical objects through spatially registered virtual annotations (Wither *et al.*, 2009). Such interfaces reduce the need to translate abstract information (e.g. encoded in maps), or switch gaze between information and physical space, as is the case with guidebooks and list-based mobile interfaces. This scenario is particularly beneficial for time-pressured visitors to unfamiliar locations (Yovcheva *et al.*, 2014).

Zimmerman (2013) argues that systems theory, communications theory, cybernetics, artificial intelligence and computer science emerging well before electronic computers, helped create the 'information revolution', and information networks no longer take the form of vast card catalogs or webs of pneumatic tubes. He argues that digital networks are flexible and organic, and when information is put at play, game-like experiences replace linear media. Media and culture in the Ludic Century is increasingly systemic, modular, customisable and participatory. Games embody all these characteristics in a very direct sense.

Increasingly, the ways that people spend their leisure time and consume art, design and entertainment will be games – or experiences very much like games.

Throughout history, many people have championed the use of play, games and game-inspired design to improve the human condition. In the mid-2000s, the confluence of web technologies, digital business models, and online and location-based gaming gave rise to the most recent manifestation of this basic idea. Mobile applications such as foursquare and websites like Stack Overflow borrowed design elements such as point scores, badges or leaderboards from social network games and meta-gaming systems like Xbox Live to motivate user activity (Nacke & Deterding, 2017). This industry practice quickly became known as gamification, which can be defined as the use of game design elements in non-game contexts (Deterding *et al.*, 2011).

Lucassen and Jansen (2014) state that in the past couple of years, gamification concepts have been successfully applied to a variety of fields such as consumer products (Deterding, 2012), education (Khan Academy, 2006), idea competitions (Witt *et al.*, 2011), citizen science (Khatib *et al.*, 2011) and marketing (Zichermann & Linder, 2010). The goals of these applications include increasing engagement, loyalty, brand awareness, motivation, purpose and/or ownership over tasks (Muntean, 2011). In their study, Lucassen and Jansen (2014) provide a detailed overview of the contemporary attitude of marketing executives towards gamification. Marketing executives' responses were positive for 9 out of 10 interview questions and they expect to adopt gamification more frequently in their future work, indicating a promising future for the gamification industry. A reason for the overwhelming positive responses is the attitude of marketing executives towards gamification as a label for a set of mechanisms that marketeers have been applying for decades.

Thus, tourist destinations should understand tourist behaviour, especially the new generation's behaviour, as an opportunity to make destinations and tourism companies more competitive, creating memorable experiences associated with technology.

Geocaching

The success of a tourist destination depends on how tourists experience the place. Many destinations seek to present genuine experiences that involve the tourist in an oriented action, combining discovery and pleasure dimensions associated with physical and mental well-being.

The newly established (but less) studied market segment is selfie-gaze tourists who travel because their peers have visited and shared it on social media, and so they also wish to take a selfie and post it on social media (Sigala, 2018a); they all want to experience it.

These creations and the information provided serve to reinforce the motivation to choose a destination, and to increase tourist satisfaction and consequently contribute to the improvement of the reputation and destination image. Earth, hiding and technology are the words that embody 'geocaching' as an excellent and genuine experience that involve the tourist in an oriented action.

According to Geocaching.com, the main features of geocaching are grouped under three concepts: the treasure hunt, the itinerancy and the game. Geocaching.com state that geocaching is the practice of hiding a container in a particular location, then publishing the latitude and longitude coordinates of the location on a geocaching website for other geocachers to find using a GPS device. In essence it is a kind a GPS-enabled treasure hunt. There are several examples of what was, and what could be, accomplished to develop alliances between geocaching and tourism. Boulaire and Hervé (2012) argue that geocaching can be considered as a type of new technologically assisted itinerancy and can also be presented as a new way to discover as well as to enhance a place, country, region or

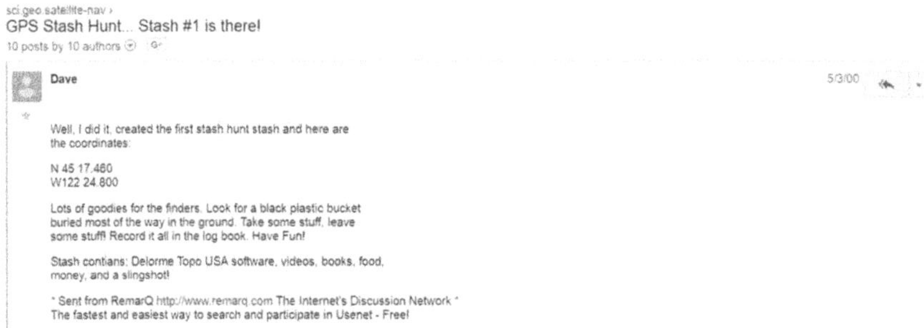
sci.geo.satellite-nav ›
GPS Stash Hunt... Stash #1 is there!
10 posts by 10 authors

Dave 5/3/00

Well, I did it, created the first stash hunt stash and here are the coordinates:

N 45 17.460
W122 24.800

Lots of goodies for the finders. Look for a black plastic bucket buried most of the way in the ground. Take some stuff, leave some stuff! Record it all in the log book. Have Fun!

Stash contians: Delorme Topo USA software, videos, books, food, money, and a slingshot!

* Sent from RemarQ http://www.remarq.com The Internet's Discussion Network *
The fastest and easiest way to search and participate in Usenet - Free!

Figure 8.2 David Ulmer's post, 3 March 2000
Source: sci.geo.satellite-nav, 13 June 2008

city. Geocachers rely on location-based technologies, such as GPS signals that are playing a more and more significant role in the consumers' everyday lives owing to their integration into many tools (e.g. smartphones, Personal Digital Assistant (PDAs), GPS and so on). The tourist also counts increasingly on these tools to identify or locate services, objects and so on. The universe of location-based service (LBS) technologies is large and diverse. Owing to the development of Web 2.0 (O'Reilly & Battelle, 2005), digital mapping has taken a prominent place in the interest in developing geocaching.

Farman (2012) explains that geocaching, as a social game, started in 2000 in the USA when GPS Selective Access was removed. People began testing the accuracy of their GPS receivers (GPSr), including Dave Ulmer who placed a container in the Portland, Oregon area and logged its coordinates onto a Usenet site (Figure 8.2). The container was found by users of the site, who logged their visits both in the container's logbook and online. Thus began the GPS treasure hunt game called geocaching. This locative game, a category with a variety of labels such as 'urban game', 'pervasive game', 'ubiquitous game', 'hybrid game' or 'flash mob game', serves as a strong example of the way that such games are changing user's relationships.

The challenge for an individual is to find a signposted spot on a map that they have set out as their destination goal, to prove that they were there and to report their 'adventure' using tourist routes, nature reserves, and historical and cultural heritage zones. Sometimes this involves crossing secular trails with breathtaking landscapes, illustrated with photographs and films that end up being shared, and are themselves strong elements of tourist attraction. These elements turn out to be, for potential tourists, inputs sufficiently capable of influencing a tourist to choose one destination over another.

The act of challenging nature, not only through contemplation but by using physical, intellectual, behavioural and technological capabilities, is something that today attracts an increasing number of people, including tourists, who achieve this desire through the practice of geocaching. These people begin to look for destinations that provide routes in advance, with waypoints placed along tourist routes. This reinforces their interest in visiting these routes because they are another challenge besides the experience of travelling to that destination. They are people, generally with a high socioeconomic status, who adhere to group initiatives related to the theme, who buy magazines and specific television channels, and who go to trade shows, are attentive to the new technologies and who mainly want to have fun.

According Geocaching.com, in April 2018 the number of active geocaches reached 3 million. In 2017, the 7 million community of geocachers hid one geocache every minute of every day, and more that 5 million 'found it' and 'attended' have been recorded in 220 different countries since the game started in 2000 (including in Antarctica).

Understanding of the successful determinants in creating competitive market niches is of the utmost importance for mature 'in-between' destinations (Machado & Almeida, 2014).

Geocaching in Madeira Island, Portugal

During recent years, the number of caches in Madeira Island grew and more than 1200 caches can now be found on the island.

In December 2016, 'Senhor da Pedra' a place located in Madeira at N32°50.110W 016°54.419 was the 'World Geocache of the Week' (Figure 8.3). One comment posted on Geocaching.com said,

> off the southwest coast of Portugal and northwest coast of Africa lies the beautiful archipelago of Madeira. It is the outermost region of the European Union and an autonomous region of Portugal. Known for its idyllic climate, flowing wine, romantic beaches, and sheer cliffs, Madeira is also home to close to 1,200 active geocaches.

They go on to say,

> You must be thinking, *Geocaching* in gloriously warm climate with stunning landscapes and flowing wine – where's the grit and the character-building adventure of *that*?! Luckily, this traditional cache provides a mildly challenging hike with a big pay off in a phenomenal location. Perfect!

Classic mature destinations, such as Madeira, should increase the number of tourists by assessing the real prospects of developing new and genuine experiences that involve the tourist in an oriented action. In this way, they can combine discovery and pleasure dimensions associated with physical and mental well-being. There is an obvious interest, shared by many classic destinations, in recovering lost ground in historical markets.

Figure 8.3 Local caches (Funchal, Madeira Island 20 June cache) and cache of the week, December 2016

Source: Geocaching.com 'Senhor da Pedra' a place located in Madeira

Methodology

To conceptualise the interest in geocaching and explore its potential for tourism development, we follow the methodology espoused by Boulaire and Hervé (2012). This involved a comprehensive exploration of multiple websites associated with geocaching activities on the island and investigating local experiences, which involved three different groups of tourists.

During the long period of research and learning, national and local sites, blogs, videos and social networks were searched for main keywords, their meanings and interrelations. The hundreds of comments provided evidence of the importance and potential that this activity has and will have for tourism in the coming years.

From the various sites visited, Geocaching.com, is accepted by the community as the authority on the matter and the operational basis of the whole system, since the main application download to access the caches and register contact with these is made from this site. A study conducted by the Geocaching.com website also helped to determine the top 15 reasons why thousands of respondents practise this activity. These reasons served as a basis for asking the 250 tourists involved in our study to find out what they thought about geocaching. The clear majority (95%) agreed, meaning that the study conducted by Geocaching.com, in addition to coinciding with our study, are validated.

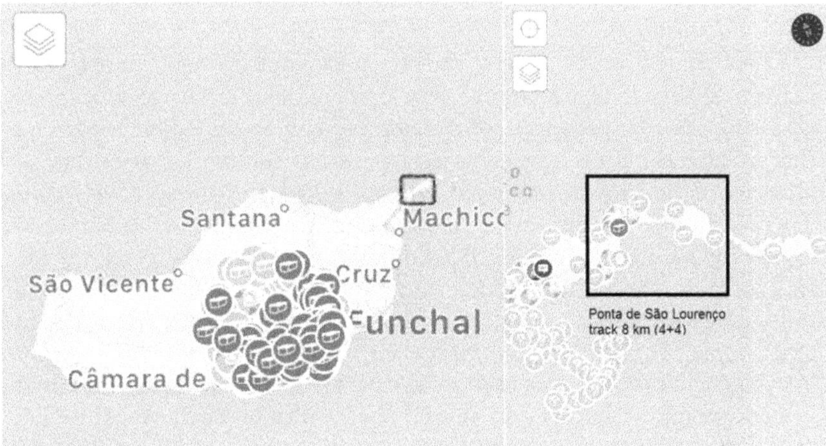

Figure 8.4 Madeira Island 'caches' and experience zone (Ponta de São Lourenço-Caniçal Machico)
Source: Geocaching.com

The survey was undertaken in March and April 2018. Tourists departing from Caniçal to Ponta de São Lourenço, (East Madeira) for a three-hour mountain walk were asked to participate in a playful activity (Figure 8.4). This included the use of a portable GPS with a pre-defined route with waypoints (check points with some information in each point about local culture, nature, geology, biodiversity and so on) and a control card. Participants were obliged to register their passage through an in-place stamp (within the cache a container with the information inside it, which participants had to find). At the end of the walk, they were asked to complete a questionnaire with some socioeconomic questions and about the geocaching activity they had just completed.

The questionnaire first collected socioeconomic information, such as age, gender, education, occupation, country of origin and salary. Regarding geocaching itself, the survey consisted of putting as questions, the same top 15 reasons why thousands of respondents practised this activity, as concluded by Geocaching.com in their study (Table 8.1) so that the tourists answered on a Likert scale (1–7 with 1 = strongly disagree and 7 = strongly agree) if each of the items corresponded to their personal interest (output in Table 8.2).

Budgetary restrictions limited the number of questionnaires to 250, assuming a response rate of 80% and a number of unusable questionnaires of 2%. The number of answered questionnaires obtained was 248, but only 246 were usable due to errors and incomplete fields. Most of the respondents were male (60%) with an average age of 46. On average, they were active, middle-class and well-educated, with a family that includes one child.

Results

This study aims to analyse the recent attempts to rejuvenate a classic mature destination based on the development of 'alternative' concepts; by assessing the real prospects of developing new and genuine experiences that involve the tourist in an oriented action, combining discovery and pleasure dimensions associated with physical and mental well-being. There is an obvious interest in investigating whether Madeira Island will eventually be included with other similar destination as offering the ideal spot for geocachers looking for a genuine experience. To assess the conditioning factors of success in this niche market, a questionnaire was applied to a sample of geocachers to assess their interest in the activity.

Geocachers.com asked their geocachers to share why they participated in geocaching activities and shared the results on their website. They stated that 'Their answers reminded everyone here at Geocaching HQ why we're more excited than ever to support this game' (see the combined results in Table 8.1).

Considering that the 'nature' element is complemented by culture and gastronomy and this is the model supported in the official strategic studies for the island tourism development, at least until 2025, combined with the fact that the age group considered ideal is signalled to be between 45 and 55 years of age, we believe that geocaching can fit perfectly into the island's tourism development strategy in the coming years. This idea can be strengthened by reading the classification of the various questions and the comments associated with each of the questions. In order to deepen the interest of the activity, other questionnaires could be carried out using the same methodology, which relate to cultural and gastronomic experiences. This could increase the dynamism of tourist activity, taking it to more remote places and at the same time, increase the satisfaction of the tourists through the discovery of unique locations, usually outside the normal tourist circuits. Even though most of the caches on the island are either on the south coast (the most touristic and most developed part of the island) or on rails or 'levadas' (a secular irrigation system that supports the most popular tourist attraction on the island), it appears that, step-by-step, new spots are emerging in less popular places, a situation that helps a greater dispersion of tourists/geo-tourists all over the island. In a simple and straightforward way, with the help of associations that dedicate themselves to the activity and even with the help of the tourists who daily 'discover' new places, there could be a greater dispersion of check points, especially in rural areas, but with a great historical, cultural and landscape interest whose tourist potential is undoubtedly something that has yet to be explored.

Discussion and Conclusion

This study aimed to analyse the recent attempts to rejuvenate a classic mature destination based on the development of 'alternative' concepts. As Madeira Island is a destination with unspoilt nature, this allows for

Table 8.1 Fifteen main reasons to be a geocacher combined with the questionnaire

Reason		Example	Mean	Standard deviation
1	You'll find that discovery, exploration and adventure still exist in the real-world.	I think it is so neat that they are hidden EVERYWHERE! There is a whole world around us that most people never notice.	5.991837	0.8962208
2	It gets you outside!	I geocache to get out of the house and enjoy nature. I have been taken to some very beautiful places in America, Ireland, and England. I would never have seen what I have seen if it wasn't for geocaching.	5.510204	0.9435382
3	It's a great way to stay healthy while having fun.	It can be great exercise and you will go places you never would of thought of, and it gives you a challenge as well!	5.416327	0.9783197
4	It's fun for the whole family.	It's the thrill of the find and it's getting the whole family out of the house and doing something fun TOGETHER! Love it!	5.284553	0.8233364
5	You can see the world.	It takes us to some amazing places off the beaten track that we would never know existed.	5.093117	1.127733
6	And experience new places like a local.	Why geocache? Because it takes you places that you never knew existed! It's such an adventure!!!	5.045082	1.078225
7	Or discover places you never knew existed right in your own neighborhood.	I discover places I've never been even though I've been there a million times.	4.639676	0.6835202
8	You can learn fun facts about planet Earth.	It's so much fun to take the family out, see nature or even places you would never have seen before and feel like you have a purpose.	4.60	0.9382579
9	And help give back too!	I geocache for fun, nature and health.	4.416327	1.260261
10	You will be delighted by the work of artists and makers.	There are some very clever cache owners that have put a lot of thought into their caches. It's more than finding just tupperware in the woods!	3.685484	0.8850255
11	And you can make the best of friends.	I REALLY like geocaching events. You get to meet and greet so many people.	3.471545	1.024796
12	You might even fall in love.	I took my wife caching on our first date to take the focus and attention off of me and put it on something I thought (and hoped) she would like. Fast forward, Linda and I are now WeFoundLove.	2.82996	1.049309
13	It will stretch you to your limits – physically and mentally.	I love a challenge and the thrill of victory.	2.715447	1.046045
14	But also guide you to moments of peace and serenity.	It's a good way for me to de-stress.	2.564516	1.424152
15	You'll have stories to last a lifetime.	Every geocacher who shared why they geocache had an incredible story to share – a story of discovery, exploration, adventure and friendship.	1.95935	1.24145

Table 8.2 Questionnaire: output data (Stata Software 16)

Variable	Observations	Mean	Standard deviation	Minimum	Maximum
Obs	248	124.5	71.73563	1	248
Country	0				
Gender	247	0.5991903	0.4910576	0	1
Age	246	46.06504	8.759592	32	63
Education	247	3.364372	0.7943051	2	5
Question 1	245	5.991837	0.8962208	4	7
Question 2	245	5.510204	0.9435382	4	7
Question 3	245	5.416327	0.9783197	3	7
Question 4	246	5.284553	.8233364	4	7
Question 5	247	5.093117	1.127733	3	7
Question 6	244	5.045082	1.078225	3	7
Question 7	247	4.639676	0.6835202	4	6
Question 8	245	4.6	0.9382579	3	7
Question 9	245	4.416327	1.260261	2	7
Question 10	248	3.685484	0.8850255	2	5
Question 11	246	3.471545	1.024796	2	6
Question 12	247	2.82996	1.049309	1	5
Question 13	246	2.715447	1.046045	1	5
Question 14	248	2.564516	1.424152	1	6
Question 15	246	1.95935	1.24145	1	5

challenges that combine adventure with good climate, landscape and the hospitality of its people. The island can offer unforgettable experiences to geocachers.

Considering that the 'nature' element is complemented by culture and gastronomy, and this is the model endorsed in the official strategic studies for the island, we have shown through the study carried out in the east side of the island that geocaching fits perfectly into the island's tourism development strategy for the coming years.

The motivation for the present research was to analyse the Ludic Century phenomenon in general and gamification in particular, as gamification has received limited research attention.

The study conducted by Geocaching.com helped us to determine the top 15 reasons why thousands of respondents practise this activity. These reasons served as a basis for asking the 250 tourists involved in our study to find out what they thought of the activity. The vast majority were unexperienced geocachers (96%) and 95% agreed with the top 15 reasons, meaning that the results of our study, in addition to coinciding with those of Geocaching.com, are validated.

Moreover, we can also say why we are more excited than ever to support this game, because 95% of the respondents were very satisfied with the ease with which they understood the operation and logic of the 'game', the ease with which they overcame the challenges of finding the way points, the lack of difficulty in progressing on the route with the help of the instructions given and above all the joy of having 'discovered' a magnificent route that they probably would never have known if they had not performed this activity.

In addition, the experience allowed us to check the adaptability of geocaching to the type of tourists that, in socioeconomic terms, are most interested in Madeira Island, given that the respondents were generally active families, middle-class, well-educated and were aged, on average, 46 years old. They are people, generally of a high socioeconomic level, who adhere to group initiatives related to the nature theme (even when not directly related to geocaching), who buy magazines and watch specific television channels, who go to the trade shows and are attentive to the latest technology and who want to have fun.

We conclude that geocaching could be a way to attract not just geocachers, but also tourists in general, by showing them unique and unfamiliar places and proposing new challenges. This can contribute to attracting the highest quality tourists and help to reinforce the destination image of the island.

Madeira Island's DMOs should include in their official web pages some specific information that will help tourists with finding geocache places, placed alongside a few photographs of geocaching experiences. They should develop applications and games in an accessible format that are compatible with popular programs such as Google Earth, Garmin-MapSource, Kompass, Instagram and so on.

The development of digital social networks apparently meant a serious step towards a more sedentary way of life. However, empirical data do not confirm these deterministic theories. Contrary to this negative perspective, the democratisation of technology proved to be a factor in spatial mobility, helping to avoid unhealthy sedentary lifestyles and even encouraging people to travel (Fernandes, 2012).

Technological advances have disrupted every tourism actor. By definition, the industry landscape and ultimately the competition are affected by disruptive innovations (Buhalis, 2003; Buhalis & Lica, 2002; Christensen, 1997; Porter, 1985), and disruptions occur when newer companies offer cheaper alternatives than those sold by established players and also, when existing markets are defined and the economic landscape reconfigured (Sigala, 2018a). While disruption often involves providing better and cheaper alternatives for customers than incumbent service providers, it is, however, unlikely that Information Technology Organizations (ITOs) with traditional business models will reduce manpower in the provision of more customised and personal experiences. ITOs have an opportunity to focus their expertise and to collaborate with platform companies that have technological expertise. They may engage in outsourcing and embrace the concept of open innovation (Hsu *et al.*, 2016).

Evidence has been given by Boulaire and Hervé (2012) that geocaching is considered to be an open object, that can be varied infinitely, notably through creative developments allowing organisations to design a framework for a consumption experience which can take into account touristic, cultural, aesthetic and/or geographic dimensions. This can be in line with the wealth of discovering a region, a locality or a place and so, in line with the development of territorial marketing such as Madeira Island.

Thus, by assessing the real prospects of developing new and genuine experiences that involve the tourist in an oriented action, the discovery and pleasure dimensions associated with physical and mental well-being can be combined. There is an obvious interest, shared by many classic destinations, in recovering lost ground in historical markets.

With regard to the policy implications, while recognising that it is not possible to control all the elements contributing to the improvement of geocaching activities in Madeira, it is possible to concentrate on niche markets, by developing specific means of advertising and promoting tourist attractions through this segment.

These results are intuitively consistent with the concept that geocaching influences visiting tourists to a destination, increasing their satisfaction and image of the destination, in this case, Madeira Island.

References

Boulaire, C. and Hervé, G. (2012) New itinerancy: The potential of geocaching for tourism. *International Journal of Management* 15 (2), 210–218.

Barabel, M. (2010) Les médias sociaux au service du marketing territorial: Une approche exploratoire. *Management et Avenir* 32 (2), 132–153.

Buhalis, D. (2003) *eTourism: Information Technology for Strategic Tourism Management.* Harlow: Prentice Hall.

Buhalis, D. and Foerste, M. (2015) SoCoMo marketing for travel and tourism: Empowering co-creation of value. *Journal of Destination Marketing & Management* 4 (3), 151–161.

Buhalis, D. and Law, R. (2008) Progress in tourism management: Twenty years on and 10 years after the internet: The state of etourism research. *Tourism Management* 29 (4), 609–623.

Buhalis, D. and Licata, M.C. (2002) The future eTourism intermediaries.*Tourism Management* 23 (3), 207–220, ISSN 0261-5177.

Christensen CM (1997) *The Innovator's Dilemma: When New Technologies Cause Great Firms to Fail.* The Management of Innovation and Change Series. Boston, MA: Harvard Business School Press.

Deterding, S. (2012) Gamification: Designing for motivation. *Interactions* 19 (2012), 14–17.

Deterding, S., Khaled, R., Nacke, L.E. and Dixon, D. (2011) Gamification: Toward a definition. CHI 2011 Gamification Workshop Proceedings, 7–12 May 2011, Vancouver, BC, Canada: ACM.

Doolin, B., Burgess, L. and Cooper, J. (2002) Evaluating the use of the web for tourism marketing: A case study from New Zealand. *Tourism Management* 23 (5), 557–561.

Egger, R. and Buhalis, D. (2008) *eTourism Case Studies: Management and Marketing Issues in eTourism*. Oxford: Butterworth-Heinemann.

Farman, J. (2012) Locative Life: Geocaching, Mobile Gaming, and Embodiment Digital Arts and Culture. In *The Proceedings of the Digital Arts and Culture Conference – After Media: Embodiment and Context*. 12–15 December 2009, Irvine, CA, USA http://escholarship.org/uc/item/507938rr.

Fernandes, J. (2012) Cadernos de geografia. Number 30/31, 2011/12 Coimbra, FLUC, 171–180.

Gillin, P. and Gillin, D. (2010) *The Joy of Geocaching*. Fresno, CA: Quill Driver Books.

Gretzel, U. and Jamal, T. (2009) Conceptualizing the creative tourist class: Technology, mobility, and tourism experiences. *Tourism Analysis* 14 (4), 471–481.

Gretzel, U., Fesenmaier, D.R., Formica, S. and O'Leary, J.T. (2006) Searching for the future: Challenges faced by destination marketing organizations. *Journal of Travel Research* 45 (2), 116–126.

Hsu, A., King, B., Wang, D. and Buhalis, D. (2016) In-destination tour products and the disrupted tourism industry: Progress and prospects. *Information Technology & Tourism* 16 (4), 413–433.

Khan Academy (2006) Khan Academy. Available at: www.khanacademy.org (accessed 17 May 2018).

Khatib, F., DiMaio, F., Cooper, S., Kazmierczyk, M., Gilski, M., Krzywda, S., Zabranska, H., Pichova, I., Thompson, J., Popovic, Z., Jaskolski, M. and Baker, D. (2011) Crystal structure of a monomeric retroviral protease solved by protein folding game players. *Nature Structural & Molecular Biology* 18, 1175–1177.

Lucassen, G. and Jansen, S. (2014) Gamification in consumer marketing – Future or fallacy? *Procedia – Social and Behavioral Sciences* 148, 194–202.

Lin, Y.S. and Huang, J.Y. (2006) Internet blogs as a tourism marketing medium: A case study. *Journal of Business Research* 59 (10–11), 1201–1205.

Machado, L.P. and Almeida, A. (2010) *Turismo: Inovação e Novas Tecnologias*. Sociedade Portuguesa de Inovação, Consultadoria Empresarial e Fomento da Inovação, Principia 2010.

Machado, L.P. and Almeida, A. (2014) The on-going process of re-invention of classic tourism destinations: The case of Nordic tourists in Madeira Island. *Scandinavian Journal of Hospitality and Tourism* 13 (Suppl. 1), 24–43.

Markets and Markets (2020) The gamification market size is projected to grow from USD 9.1 billion in 2020 to USD 30.7 billion by 2025. Available at: https://www.marketsandmarkets.com/Market-Reports/gamification-market-991.html (accessed 17 December 2020).

Muntean, C.I. (2011) Raising engagement in e-learning through gamification. Proceedings of the 6th International Conference on Virtual Learning (ICVL) 28–29 October 2011, University of Cluj-Napoca România (pp. 323–329). ACM Press.

Nacke, L.E. and Deterding, S. (2017) The maturing of gamification research. *Computers in Human Behavior* 71, 450–454.

Neuhofer, B. and Buhalis, D. (2013) Experience, cocreation and technology: Issues, challenges and trends for technology enhanced tourism experiences. In S. McCabe (ed.) *Handbook of Tourism Marketing* (pp 124–139). London: Routledge.

O'Connor, P. (2008) User-generated content and travel: A case study on Tripadvisor.com. *Information and Communication Technologies in Tourism* 2008 (2), 47–58.

O'Reilly, T. and Battelle, J. (2005) What is web 2.0? Design patterns and business models for the next generation of software. Available at: https://www.oreilly.com/pub/a/web2/archive/what-is-web-20.html (accessed 17 December 2020).

Pine, B.J. and Gilmore, J.H. (1998) Welcome to the experience economy. *Harvard Business Review* 76 (4), 97.

Porter, M.E. (1985) *Competitive Advantage: Creating and Sustaining Superior Performance*. New York: Free Press.

Ramaswamy, V. (2009) Leading the transformation to co-creation of value. *Strategy & Leadership* 37 (2), 32–37.

Seaborn, K. and Fels, D.I. (2015) Gamification in theory and action: A survey. *International Journal of Human-Computer Studies*, 74, 14–31.

Secretaria Regional da Economia Turismo e Cultura (2016), Plano Estratégico para o turismo 2017-2021

Sigala, M. (2018a) New technologies in tourism: From multi-disciplinary to anti-disciplinary advances and trajectories. *Tourism Management Perspectives* 25 (2018), 151–155.

Sigala, M. (2018b) Market formation in the sharing economy: Findings and implications from the sub-economies of the Airbnb. In S. Barile, M. Pellicano and F. Polese (eds) *Social Dynamics in a Systems Perspective New Economic Windows* (pp. 159–174). Springer: Cham. http://doi-org-443.webvpn.fjmu.edu.cn/10.1007/978-3-319-61967-5_9.

Stamboulis, Y. and Skayannis, P. (2003) Innovation strategies and technology for experience-based tourism. *Tourism Management* 24 (1), 35–43.

Tussyadiah, I., Wang, D., Jung, T.H and tom Dieck, M.C. (2017) Virtual reality, presence, and attitude change: Empirical evidence from tourism. *Tourism Management* 66, 140–154.

Wang, Y.C., Yu, Q. and Fesenmaier, D.F. (2001) Defining the virtual tourist community: Implications for tourism marketing. *Tourism Management* 23, 407–417.

Wither, J., DiVerdi, S. and Hollerer, T. (2009) Annotation in outdoor augmented reality. *Computers & Graphics* 33 (6), 679–689.

Witt, M., Scheiner, C. and Robra-Bissantz, S. (2011) Gamification of online idea competitions: Insights from an explorative case. INFORMATIK 2011: Informatikschafft Communities 41. *Jahrestagung der Gesellschaft für Informatik*, 4. Berlin.

Xu, F., Weber, J. and Buhalis, D. (2013) Gamification in tourism. In Z. Xiang and I. Tussyadiah (eds) *Information and Communication Technologies in Tourism 2014* (pp. 525–537). Wien-Springer.

Yovcheva, Z., Buhalis, D., Gatzidis, C. and van Elzakker, C. (2014) Empirical evaluation of smartphone augmented reality browsers in urban tourism destination context. *International Journal of Mobile Human Computer Interaction* 6 (2), 10–31.

Zimmerman, E. (2013) Manifesto for a Ludic Century. Available at: https://ericzimmerman.wordpress.com/2013/09/09/manifesto-for-a-ludic-century/ (accessed 11 May 2018).

Zichermann, G. and Linder, J. (2010) *Game-Based Marketing: Inspire Customer Loyalty Through Rewards, Challenges, and Contests*. Hoboken, NJ: John Wiley & Sons.

9 Advergaming in Tourism: Spanish Cases

Pablo Garrido-Pintado

Introduction

The loss of effectiveness of traditional media has led companies to seek new methods of making contact with the consumer. In new digital and interactive scenarios, the consumer 'is someone who does judge, gets engaged and in most cases dominates, the language of marketing and advertising; an expert consumer who decodes the intentions of advertising campaigns and brand strategies' (Del Pino & Galan, 2010: 57–58).

In this context, advertising is changing. While in the past, there were many repetitive messages in the mass media, now new formulas are being experimented with. For several years, that mix of entertainment and persuasion have been explored. These formulas seek to strengthen the user experience through engagement.

Towards a New Advertising Paradigm

The phenomenon of advertising saturation has been widely discussed in scientific literature (Bekerian & Baddeley, 1980; McFall, 2004; Reinares & Reinares, 2007; Rumbo, 2002). Araceli Castelló (2012: 113) defines advertising saturation as:

> Percentage of time or space allocated to advertising in relation to the recommended or legal limits, or in relation to the informative and / or informative contents of the corresponding media or support.

Advertising is changing. The model of persuasive messages directed to large masses is progressively extinguishing. We are moving towards a new advertising paradigm (Carlson, 2015; De Mooij, 2018; Viñarás *et al.*, 2015). Starting from the reflections of Méndiz (2010), we debate the fundamental axes of new advertising: the hybridisation between advertising and information, bidirectional communication and playful experience.

- Hybridisation between information, entertainment and advertising tries to attract an increasingly fragmented audience (Covaleski, 2010;

Selva, 2009), reactive to the reception of advertising messages. This situation where media and adverts lose credibility favours the expansion of hybrid formats such as native advertising (Campbell & Marks, 2015; Matteo & Dal Zotto, 2015). The success of native advertising depends on the content. This content should not have a promotional tone and should provide value to the user. For example, in the case of a tourist brand, we can talk about travel destinations. But we do not seek to sell, nor does it demand the user to convert immediately, since we understand that it is still in the early stages of the conversion funnel. The Nudge platform (Toib, 2018) collects cases of native advertising in the world of tourism; for example, those made by Tourism Ireland, Cunard, Destination Canada, Visit Copenhagen. Trivago or The Corning Museum of Glass.
- Thanks to new media, the receiver plays a more active role (Daugherty *et al.*, 2008; Khan, 2017; Lee & Hong, 2016). Internet and other digital media give more possibilities for communication with the recipients. For example, users waiting at airports or stations prefer to play with their smartphones instead of reading books or newspapers. Many companies or entities managing tourism take advantage of these moments to promote their brands using advergaming. This is the case of the city of Rotterdam with the game 'Get Lost in Rotterdam' or 'Malmö's Treasure Hunt' (Yılmaz & Coşkun, 2016).
- Brands not only seek persuasion, notoriety or memory in the mind of the consumer (Klein *et al.*, 2016; Méndiz, 2010) but propose playful experiences that link them with their audiences from an emotional point of view. For example, American Airlines launched a mobile application that allowed consulting the elite points of the users using gamification (Xu *et al.*, 2013).

New forms of leisure are emerging, as pointed out by Cañete *et al.* (2013: 1–23): 'In a digital landscape overpopulated with brands, services and products, the capacity for attraction can no longer reside only in the message, but in communication and in relationship established between user and brand'.

Faced with exhaustion of traditional media, new supports related to the internet and Web 2.0 (blogs, social networks, multimedia repositories, etc.) and services linked to mobile communication started to explode their advertising potential. In parallel, ways that mix persuasion and entertainment are probed, either through actions of ambient marketing, augmented reality or exploring the capacity of involvement in videogames.

Videogames: A Booming Industry in the Spanish Market

Spain occupies a prominent place in the European market of the videogame industry, with billing figures of €1163 million (Anuario de la

Industria del Videojuego [AEVI], 2016). Of the total, €781 million comes from shop sales and €382 million through internet. According to data published by the National Telecommunications Observatory (ONTSI, 2018), 43.3% of individuals over 15 years old have a videogame console (fixed or portable). In addition, 53.9% affirm to have downloaded games on their smartphone or tablet, and more than half of them play several times a week. According to PWC research (PWC, 2017), the PC game market is considerably smaller than that of consoles. On the other hand, the rapid growth of smartphones in Spain (up to 35.2 million in 2016) has helped social/casual games to take a 47.7% share of the market. It is a high level of participation, compared with surrounding countries. During 2016, the impact caused by Nintendo's arrival in the mobile market, with conventional products such as Super Mario Run or the enlarged reality game of Pokémon Go, was remarkable.

According to data from AEVI (2016), in Spain there are 15 million players. The highest percentage of use is concentrated in young users; the percentage of use of videogames in children and adolescents is above 65%. However, as the age increases, these percentages decline. Between 25 and 34 years old, the percentage of players is 46%, a figure that drops 10 points between 35 and 45 years old. In Spain, the average weekly time dedicated to videogames is 6.2 hours.

According to Eduardo Madinaveitia (2010: 44), 'the creation of videogames is already a very important economic activity, which moves a quantity of money superior to that produced jointly by music business and cinematographic production'.

The study Entertainment and Media Outlook (2017–2021) prepared by Price Waterhouse Coopers (PWC, 2017: 28) points out a series of trends that favour the development of online games. The need to 'focus efforts on building businesses and brands anchored in communities of active and high-value fans, united by common interests, values and passions'. We talk about the fan-centric model (Shih *et al.*, 2014): this is a model that involves the construction of brands that drive loyalty in a very competitive environment, increasing the customer's lifetime. To achieve this, it is necessary to create a comprehensive user experience with impeccable customer service and personalisation of products.

This global trend has caught the attention of advertising. Thus, as Sebastián and Carcelén (2014: 250) point out:

> Video games have become one of the platforms most used by advertisers who are looking for new formulas to spread their messages effectively, something that is logical considering the effectiveness indexes that reflect a greater acceptance among users.

A study conducted by Steve Bellman and colleagues (2014) of the University of Murdoch, Australia, determines that the quality of interaction between user and the videogame favours the memory. The interactivity of videogames allows the users to immerse themselves in the story and forget what happens around them. According to Bellman *et al.* (2014: 277) 'those people who enjoy an advertising experience through an advergame or computer game, have a greater ability to remember the products'.

Videogames and advertising

These capabilities enable advertising to take advantage of the possibilities offered by videogames. According to the White Book edited by the Spanish Association of Production Companies and Developers of Videogames and Entertainment Software (DEV, 2014), videogames may include advertising for which the owner of the game earns income.

The alternatives are:

- In game advertising: games that contain advertising. Good announcements (e.g. billboards in a car race or at a football game) or products that are part of the action (product placement). Very striking is the example of Edreams, an online travel agency, which personalised several levels of the Little Big Planet game.
- Around games advertising: habitual practice in online games. Advertising that surrounds the game and usually appears before or after the game. For example, Turkish Airlines sponsor of the Euroleague Basketball, includes video ads and mini-games before accessing the reward screens of the game Asphalt 9.
- Advergaming: a brand is specifically promoted throughout the game, because it is the one that finances the development of the game. For example, Dubai Tourism commissioned Gameloft in 2016 to develop a game in which users could enjoy virtually walking through its dunes, jumping by parachute or kitesurfing on the coasts of the country.

Advergaming is an advertising strategy that mixes entertainment and advertising with the aim of approaching the consumer with a playful and non-intrusive approach. As a consequence of the fragmentation of the audience (Couldry & Turow, 2014; Fletcher & Nielsen, 2017; Moriarty *et al.*, 2014), it raises an alternative way to attract the attention of the public.

The term advergaming consists of two words: advertising and gaming. According to Sharma (2014: 248) 'advergaming is the process in which advertising message, logo of company or other information about the brand is incorporated with online or offline videos games'. More neatly, Chen and Ringel (2001) consider this phenomenon as the 'use of interactive gaming technology to disseminate to consumers integrated advertising messages...through the incorporation of branding in the game environment'. Klein *et al.* (2016: 5761) state 'the main objectives of this technique are to build brand awareness, offer product information, drive traffic to web sites and provide a means to compare similar products'.

Advertising agencies, advertisers and development companies take advantage of this communication technique to achieve their marketing and advertising objectives. For Purswani (2010), advergames are beneficial to strengthen brand recognition through interactivity and the value of repetitive training. However, authors such as Lee and Cho (2017: 235) point out that this is a covert marketing technique, trying to promote their characteristics as 'interesting', 'adventurous' and/or 'exciting'. Nevertheless, they continue, 'emphasizing these attributes no longer helps any advergame to differentiate itself from its competitors...thus losing its advertising power'.

Authors such as Selva (2009) and Shan *et al.* (2016) highlight that advergames are advertising strategies of the pull type, where brands attract consumers to their message, creating a pleasant space where they can stay as long as they want. Users do not face merely promotional messages. The receiver consumes the image of the brand, while enjoying the videogame. The user interacts with the brand on their own initiative, participates because they perceive a value and freely decide the degree of relationship they establish. For example, Caldea Health Resort, located in the Principality of Andorra, launched the game 'Desestressing' in 2008. The players, once registered, competed to win one of the 100 free stays at the spa. The tool, integrated into the rest of the company's marketing actions, aimed to publicise the attractions of the leisure centre, transmit brand values, generate traffic to the online channel and increase the number of online bookings.

The commercial component establishes the main difference between advergames and gamification. An advergame is created on purpose to promote a product or brand. The concept of gamification is to apply game dynamics in different areas, managing to enhance the motivation, concentration and effort of people in the activities (Groh, 2012).

The novelty of the term and the tremendous dynamism of the advertising sector when it generates new formats does not allow for establishing a unified typology of advergames. Chen and Ringel (2001) define three types of integration between advertising and game: associative (establishes linkage between the product, lifestyle and activity shown in the game), illustrative (where the product plays a leading role) and demonstrative (the user interacts and experiments with the product in the game universe).

Michelle Nelson (2005), in her work 'Exploring consumer response to advergaming' establishes a rich taxonomy that details the different uses of brands in videogames (Table 9.1). The classification proposed by Nelson refers to examples of advertising inserted in the game (in game advertising). Only in the case where the brand constitutes a main part of the game, would we be talking about advergaming.

The phenomenon has received attention from the scientific community. Since the first study by Chen and Ringel (2001) many investigations have been published. Some of them focused on the definition of the term

Table 9.1 Advertising taxonomy: Identifying brand uses in games with some examples in the tourism industry

Brand use	Proposed brand effectiveness	Examples
Sponsorship/banner ads	Brand awareness, reinforce brand image Enhance consumer involvement	FIFA, NBA Live; Pizza Hut in Crazy Taxi
Contests, gamelettes on websites	Increase stickiness of site, involve consumers	Brasil Quest; A weekend in Lyon (France); Jack across the sea for Marina Suites in Canary Islands; Destination honeymoon, a contest from Airbnb.com
Background visual ads and brand props (e.g. billboards, side-boards, clothing accessories)	Brand recall Familiarity may enhance brand attitudes	Verizon and Energizer billboards in Ana Wake's videogame; Subway in Uncharted 3
Background (auditory). Licensed music, brand sound effects, sports commentators	Increases game realism and consumer involvement, good method of cross-selling	PSC, FIFA
Brands are major parts of game play (equipment tools)	Consumer can try products in virtual life before buying products in real life. Build and reinforce brand loyalty	Windows phone in Quantum Break; Doritos and AXW in 'Metal Gear Peace Walker'; Mercedes Benz in Mario Kart
Characters are branded images (real or fictitious)	Brand liking relationships. Aids in consumer identification with character brand	Burger King in 'Fight Night Round 3'
Games players create their own ads brand through customisation	Increases brand involvement loyalty	Nascar Racing, FIFA,
Augmented reality games	Use of augmented reality to show the characteristics of a product/brand	'Fall of the Wall' (Berlin) by Tripventure; Stockholm Sounds; 'Fairy and the Safe' mobile app from Eftelin Theme Park in the Netherlands
Complete advergaming	Game specially designed for a brand	Cazamaletas from Lastminute.com; Desestressing from Caldea Health Resort; The game of Paella (Valencia); Look me in Benidorm; Smile Land in Thailand, Run Island (Reunion Island)

Source: Based on Nelson (2005)

(Buckner *et al.*, 2002; Marolf, 2007), on consumer response (Nelson, 2005; Gurău, 2008; Lambert & Lee, 2018; Svahn, 2005; Van Reijmersdal *et al.*, 2015; Vashisht & Royne, 2016) or on its effects (Chaney, 2004; Dobrow, 2004; Yang *et al.*, 2006). In the case of Spain, publications have focused on the definition of the concept (Haro, 2002; Martí-Parreño, 2008; Méndiz, 2010; Sánchez, 2006; Selva, 2009) and in the applications in different sectors (Del Moral, 2014; González & Francés, 2009).

However, the real investment is far below the expectations created around this phenomenon. In the Infoadex study of the Advertising Investment 2017, the advergaming appears described in the area of mobile marketing. Neither in Infoadex nor in the latest studies of the IAB (Interactive Advertising Bureau) there are investment figures for this format, so it is difficult to make a real calculation of the money invested in this type of actions.

This lack of measuring instruments clashes with the positive vision of the academic world. For Santos *et al.* (2007) and Lee and Cho (2017), the benefits of an advergame are as follows: reinforcement of the brand image, advertising that does not intrude, generates involvement and improves purchase intention, increase in traffic and time spent on the website that hosts the game, creation of databases and virality and generating brand recommendations through word of mouth. For Purswani (2010) an advergame provides interactivity and replayability. Normally, an advergame encourages users to return; generating engagement through achievements and unlocks.

In recent years, the progressive growth of the smartphone market has led to the start-up of gamification initiatives in the mobile market. For Çeltek (2010), mobile advertising is no different from traditional advertising.

The Gamification of Tourism

The tourism industry is a pioneer in the adoption of technological innovations (Buhalis & Law, 2008). It began in the 1960s with the adoption of electronic data transfer systems and, later, with the arrival of the internet. Xu *et al.* (2013: 256) point out that gamification was present in tourism even before the definition of the concept itself. For these authors 'every loyalty card or frequent flight program has a gamification approach'. These programs, born in the 1980s in the Anglo-Saxon world (Klophaus, 2005), allow for the accumulation of points with trips, gasoline/petrol refuelling and purchases, and then exchange them for gifts or free tickets and accommodation.

However, the development of communication strategies that incorporate games is relatively recent. The first initiatives arouse around 2006, although the strongest impetus has occurred in the past five years. The gamification environment created by Google, with its Google Local Guides product, is noteworthy. Through this platform, the technological giant encourages its users to feed their maps with content, adding sites, editing information, contributing photographs or reviews. Each contribution is awarded with points and the most active users advance in level.

Each level has a number of advantages (receive an informative newsletter, get badges, invitations for events, free hosting on Google Drive, etc.).

Xu *et al.* (2017) suggest that the benefits of tourism gamification include internal and external aspects. On the one hand, the internal function is linked to the use of games for the training of the staff; on the other hand, the external function refers to the experiences of tourists before, during and after the trip:

- Before the trip: games that feed the tourist information and inspire the visitor. Interest is awakened by making a trip to a specific destination or establishment. Sigala (2015) points out that tourism companies' main objective is to provide valuable information in order to activate the need to travel to a place.
- During the trip: games as an entertainment tool that improve on-site experiences. For example, Buhalis and Foerste (2015) in the paper, 'SoCoMo marketing for taveling and tourism', mention the example of Marta. Through her mobile application that integrates social networks in a specific context, during a visit to an art gallery, she publishes positive updates in real time about her experiences and connects them with photos of the museum. The application connects with friends or relatives who also are making the visit.
- After the trip: share experiences through games. Rewards systems can encourage repeat visits. In addition, visitors can document the information obtained in a personal blog or on a review site, which can influence friends or other potential visitors of a certain destination (Buhalis & Foerste, 2013).

Weber (2014) points out that several applications are used in the three phases of the trip: virtual reality before the trip, during the trip: augmented reality games, gamified tourist guides, immersive experiences in cultural heritage, transmedia storytelling and gamified experiences in accommodation as well as exchange of information on social networks after the trip.

Methodology

The present work aims to make an approximation of the phenomenon of advergaming in the Spanish tourism industry. The methodology used for this research focuses on the review of existing literature, documentary analysis and the study of examples and relevant cases. The case study is a qualitative research methodology that provides empirical data that allow analysis of a phenomenon determined, drawing conclusions and facilitating the understanding of theoretical constructs or complex arguments (Gummesson, 2007). The strategies of gamification and advergames that were started, either for the launching of specific actions or as a fundamental part of a comprehensive communication strategy will be analysed.

Table 9.2 The analytical framework

Name, title and brief description	**Brief presentation of the game**
Type of game and technical characteristics	Description of the dynamics of the game: geolocation, visual search, treasure hunt, prizes, check-in places, quiz, contest. Description of the technical characteristics of the game and type of platform used
Who is it for?	The target audience of the game is described
Release date	Date of presentation/launch of the game and updates, if applicable
Objectives	Main objectives of the game. Relationship between the dynamics and market objectives
Solution	Detailed description of the game, explaining in detail on which devices it works, the mechanics of the game, prizes and the possibilities for promotion of the attraction/destination
Results	Download data of the tool/application are offered. Although, owing to privacy policies, many of the technology providers omit this valuable information

The data were collected through the internet between September and December 2018. To search for cases, snowball sampling was used. According to Çeltek (2010: 270), snowballing 'is a technique for developing a research sample where existing study subjects recruit future subjects from among their acquaintances...thus, the sample group appears to grow like a rolling snowball'.

This methodological technique is often used in hidden populations that researchers find difficult to access. Finally, 10 cases of advergames and applications for mobile phones and tablets with intensive use of gamification as a resource were found to enrich the tourist experience. In the selected sample (see Table 9.2), we found videogames and some integrated experiences (web, apps, virtual reality, augmented reality, 360°) that included many types of gamification (treasure hunting, testing contests, registration points, visual search and geolocation routes). This variety of cases from different platforms means that gamification includes not only videogames, but also mobile applications that use gamification mechanisms (Xu *et al.*, 2017). There are more cases of gamified mobile applications because their development cost is less than that of a videogame.

Results

The following case studies illustrate the most relevant examples of the use of advergaming and gamification strategies implemented by companies and institutions related to tourism are discussed.

Case study 1: Amazing City Castellon

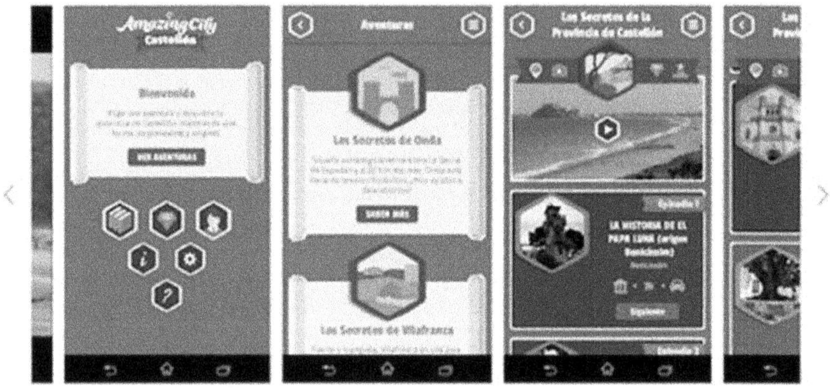

Figure 9.1 Amazing City Castellón: Screenshots of Amazing City Castellon
Source: Amazing UP – Innovación para el turismo

Table 9.3 Amazing City Castellon

Name, title and brief description: Amazing City Castellón is a gamified mobile application that offers the possibility of winning prizes in hotels and restaurants. It offers more than 100 adventures that take place throughout the province of Castellón. By overcoming tests, you can discover the secrets of places such as Peñíscola, Morella, Benicassim, Cabanes, Ribesalbes or Onda.

Type of game and technical characteristics: Application for mobile phones. Augmented reality game/treasure hunt: searching and photographing elements/quiz: answering questions about places/share on social networks. Available on Google Play and the App Store. Combination of augmented reality and visual search. It allows capturing of QR codes.

Who is it for? For all ages.

Release date: October 2014/last upgrade on November 2016.

Developer: Amazing City was developed by Inoibérica – currently Amazing Up – with the collaboration of University Jaume I, Grupo Zeta and Mago Yunke.

Objectives: The objective of the application is to prevent the tourist from paying a simple visit to places of interest. Instead, the tourist lives an authentic adventure linked to the history, gastronomy or culture depending on the location of the province.

Solution: Once the tourist downloads the application onto their mobile phone, they can start completing a series of adventures in the different municipalities, which they can choose according to their interests, tastes or even time availability. The application guides the tourist throughout the municipality and asks them to carry out different tasks in which the mobile phone acts as an interface and in which different cutting-edge technologies such as augmented reality or a visual search are combined. Depending on the player's ability to solve the different tests that the application presents on site, the system will give points as well as access to new games or content of tourist interest. The application includes an important list of awards related to hospitality and catering companies in the province, which have seen in this tool an opportunity to promote their establishments by linking their tourism offer to mobile devices.

Results: More than 5000 downloads on the Google Play platform.

Case study 2: Imageen Tarraco

Figure 9.2 Imageen Tarraco
Source: Google Play

Table 9.4 Imageen Tarraco

Name, title and brief description: Application that combines augmented reality and virtual reality. Allows visitors to visit Tarragona and Costa Daurada known as the ancient Roman city of Tarraco.
Type of game and technical characteristics: Augmented reality and virtual reality game. The viewing is activated by capturing codes in the city. Optimised application for smartphones, tablet, PC and virtual devices (Gear VR Samsung, Oculus Go and Google Cardboard).
Who is it for? Tourists who visit the city of Tarragona.
Release date: March 2016.
Developer: Imageen. Reliving History.
Objectives: Allowing the visitor to become immersed in the Roman past of Tarragona.
Solution: The smartphone and the tablet are 'windows to the past' through which it is not only possible to admire the grandiosity and original layout of the monuments and ancient spaces that are being visited, but it can also help to immerse the tourist in the atmosphere of this splendid era of the city. The tourist can enjoy the four most important monuments of the Roman Tarraco which date to the second century CE, when the city was the capital of the Hispanic province of Tarraconensis. The visit includes panoramic views in 340º mixing the past and the present and follows the steps of a virtual guide who explains the provincial forum.
Results: More than 10,000 downloads on the Google Play platform.

Case study 3: Haro in Game

Figure 9.3 (a) Haro in Game

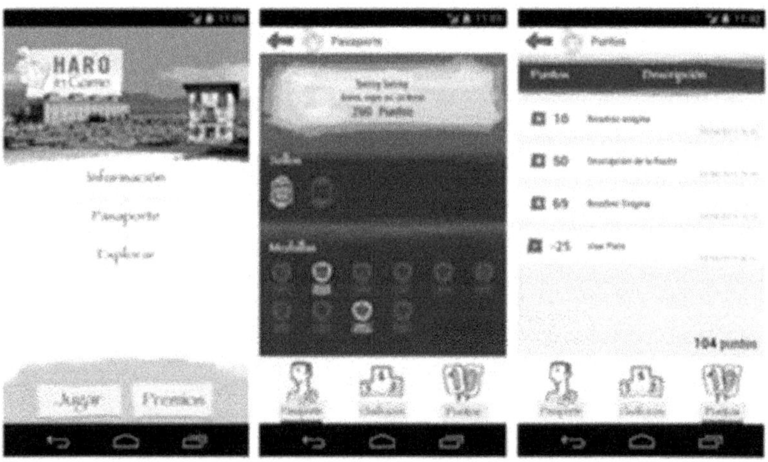

Figure 9.3 (b) Haro in Game

Table 9.5 Haro in Game

Name, title and brief description: The online travel agency Rumbo and La Rioja Turismo developed the mobile application 'Haro in game'. This interactive application was presented at the same time as a virtual tour through the exhibition 'La Rioja Open Land'. The exhibition was celebrated in the city of Haro (La Rioja, Spain) from 22 March to 13 October 2013.
Type of game and technical characteristics: Virtual route with geolocation elements. Integrated with social networks such as Foursquare, Facebook, Twitter and Instagram. It was available on Google Play and the App Store until the end of 2014.
Who is it for? For all ages.
Release date: 2013 – the application was developed as a virtual tour for the exhibition, La Rioja Open Land. Once the exhibition was over, the app continued offering a route round the city of Haro, informing tourists about its gastronomic, tourist and cultural resources. Currently it is not possible to download the app from any mobile application platform.
Developer: Technical department of Rumbo.es, an online travel agency.
Objectives: Make La Rioja known to internet users and attract travellers to the region.
Solution: Geolocalised route through Haro (La Rioja) that informed the tourist about the gastronomic and cultural resources of the city. With a design based on geolocation, it allowed discovering Haro through clues and puzzles that were found throughout the city and a 'game' in which the participants accumulated points as they passed tests. Within the game the participants with the highest score were awarded with a weekend in Haro. The application was integrated with Foursquare, Facebook, Twitter and Instagram and could be downloaded for Android and IOS.
Results: No data on interaction and number of downloads are available.

Case study 4: Marco Topo in the Museum

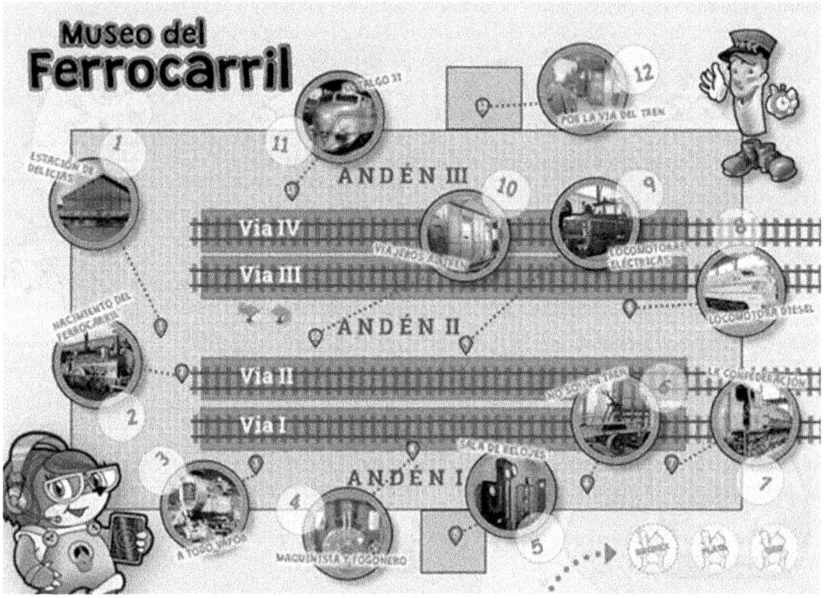

Figure 9.4 Marco Topo in the museum: map with the Marco Topo tour in the museum
Source: Marco Topo / Museo del Ferrocarril

Table 9.6 Marco Topo

Name, title and brief description: Marco Topo is a game that makes a visit to different museums interactive for children.
Type of game and technical characteristics: Game is on a web platform. Accessed through a code.
Who is it for? Children
Release date: 2016.
Developer: Marco Topo SL.
Objectives: To make museums visits more interactive.
Solution: Upon arriving at the museum, the children receive a tablet and a board where they have to attach stickers of the '*flopis*'. The flopis are characters of Marco Topo that appear as the children are solving the challenges posed by the game that appears on their screens. While the parents look at the paintings/attractions of the museum, the children go at their own pace. This game is sold at the ticket offices and in the stores of the museums.
Results: The creators have developed more than 30 tours for different museum spaces such as the Sorolla Museum and the Museum of Natural Sciences of Madrid. Recently, they have incorporated spaces such as the Museum of Prehistory and Archaeology of Cantabria (MVPAC), the Railway Museum of Madrid, the Museum of Human Evolution of Burgos, Pagos del Rey Wine Museum and the tourist routes of Estepona (Málaga), Cabra (Córdoba), Bilbao (Vizcaya), Vitoria (Álava) and the Historic Botanical Garden of La Concepción in Málaga. According to data from the developer (M, Vila, personal communication, 27 December 2018), more than 10,000 families have played one of the Marco Topo games. The average game time is between 60 and 90 minutes. The success rate (completed items) is 80% in museums and 60% in cities.

Case study 5: The compass of Ili Palmir in Elche (Alicante)

Figure 9.5 (a,b) Ili Palmir

Table 9.7 The compass of Ili Palmir

Name, title and brief description: The compass of Ili Palmir is a tourist game that is aimed at the smallest visitors of Elche. On this platform, the user becomes the adventurer Ili Palmir and thanks to his mission, he will explore the city, through overcoming 27 challenges in order to advance through the game.
Type of game and technical characteristics: Virtual route through the city. Mobile app available on Google Play and in the App Store.
Who is it for? Children.
Release date: 2016.
Developer: GorilAPP.
Objectives: To explore the city of Elche through several adventures.
Solution: In this app, the character of Ili Palmir will guide his little adventurer with a compass, via a mobile phone, showing where they have to go to reach each of the objectives with the map of the city of Elche in which the complete the route is shown. To make the game more interactive and entertaining, the application uses different illustrations and alarms in the different phases of the adventure, showing the history of the places and monuments that are being seen at that moment. The user can share their achievements in social networks and obtain a diploma, which appears on the screen of their device, certifying that they have successfully carried out the mission of Ili Palmir. In addition, the app gives the option to complement this experience with a gymkhana in which different questions related to the visit are raised, which contributes a degree of learning about the city to the little adventurers.
Results: According to Google Play data, the app has been installed more than 1000 times.

Case study 6: Borgia Territory

Figure 9.6 Territorio Borgia
Source: Play & Go Experience

Table 9.8 Territorio Borgia

Name, title and brief description: This application presents the history and legend of the Borgia family and its main characters (Callistus III, Alexander VI, Benedict XIII and Clement VIII), as well as the patrimonial elements of the Valencian community directly related to them.
Type of game and technical characteristics: Application with geolocated information. Users must complete a series of tests to obtain rewards. The game presents images, videos, visual searches of heritage elements, 2D maps created with Mapbox and topographic bases in 3D developed with MapZen. It offers 360° images of the main spaces, as well as geopositioning and augmented reality of the main resources.
Who is it for? For all ages.
Release date: 2018.
Developer: Play&Go Experience.
Objectives: The objective is to offer a vision of the territory that favours an all-year round demand and positions the Valencian community as offering a cultural vision of the rest of the tourism products.
Solution: Through the application users discover the origins of the Borgia family and its first pope, Callistus III; discover Alexander VI, the second pope, and the legacy of Francisco de Borgia and other notable members of the family. The story's thread allows users to collect Borgian maravedis, which appear as they travel through the territory and which later will be useful for other games. The mission of the game is to collect the gems in the Borgia territory, visiting the main tourist resources, until completing the 10 gems of Alexander VI. For this, users have the help of a narrator, Alex, a robot 'alter ego' of Pope Alexander VI, who will take users back in time to 500 years ago and will accompany them throughout the adventure.
Results: According to Google Play data, the app has been installed more than 100 times.

Case study 7: Jai&GO, Aste Nagusia guide (Bilbao)

Figure 9.7 Aste Nagusia
Source: Play & Go Experience

Table 9.9 Jai&GO

Name, title and brief description: Jai&Go is a guide to learn about all the points of interest and most relevant events of the Aste Nagusia/Semana Grande de Bilbao 2018.

Type of game and technical characteristics: Application with geolocated information. Includes gamification mechanics, geolocation, image recognition, augmented reality and panoramic images. Mobile app available on Google Play and in the App Store.

Who is it for? For all ages.

Release date: 2018.

Developer: Play&Go.

Objectives: To show the popular, participative and inclusive character of the Bilbao festivities. It combines a model that mixes leisure and culture through new technologies.

Solution: To complete the different objectives and missions, the player must go to different locations in their environment along the chosen path and can:

- use the items that have been collected previously, in an augmented reality test via a mobile phone;
- take a selfie and share it on social networks;
- rate and comment on the places visited;
- check in at a certain location;
- respond to a quiz or survey; and
- recognise and discover different images, places, treasures, etc.

Results of the campaign: According to Google Play data, the app has had more than 1000 installations.

Case Study 8: App Teruel Go

Figure 9.8 (a,b) Teruel Go

Source: iTunes Store /Imascono/Aragon Government

Table 9.10 Teruel Go

Name, title and brief description: Gamified app to learn about the modernist buildings of the city of Teruel.
Type of game and technical characteristics: Gamified application with geolocated information. Mobile app available on Google Play and in the App Store.
Who is it for? Tourists visiting the city of Teruel.
Release date: 2018.
Developer: Aragon Turismo, Ittainnova and Imascono.
Objectives: Tourism promotion and enriching the travel experience.
Solution: The application enables learning about the most emblematic buildings of Teruel through a personalised tour. To discover the secrets of Teruel's modernism, tourists will have to reach the points of interest indicated on the map and open the digital locks under which the information is enclosed. The application presents a game in which the tourist has to complete the points at each visit, is able to take photos and learn about other points of interest in the city.
Results: According to Google Play data, the app has had more than 1000 installations.

Case study 9: Video contest for the tourist promotion of the province of Seville

Figure 9.9 Sevilla

Source: Prodetur, S.A.U./Turismo de la Provincia de Sevilla

Table 9.11 Seville

Name, title and brief description: 'There is another Sevilla...full of experiences'. With this motto, the deputation of Seville looked for tourists by capturing the essence of the different localities of the province.
Type of game and technical characteristics Video contest/web platform.
Release date: 2014.
Developer: n/d.
Objectives: To promote the heritage, culture and landscapes of the province of Seville and encourage audience participation.
Solution: A contest to elaborate on a promotional video of the province of Seville. In the first place, the interested participants had to record a video with a mobile phone or camera. Subsequently, they had to access the website or the Facebook profile of the institution. They had to use some of the eight proposed hashtags: #fiestas, #gastronomy, #patrimonio (heritage), etc. Finally, the audience voted for their favourite video. The campaign, disseminated through social networks, had a series of incentives for the winners: a panoramic flight through the province, a visit to a cattle ranch, gastronomic experiences and accommodation in charming hotels and so on.
Results: n/d.

Case study 10: 'Enigma Galdiano': Free game of augmented reality

Figure 9.10 (a,b) Galdiano
Source: Museo Lázaro Galdiano

Table 9.12 Enigma Galdiano

Name, title and brief description: 'Enigma Galdiano' is an innovative game that aims to turn Madrid's museums into a fun and educational experience for families. In the first phase, it is located in the Lázaro Galdiano Museum but the idea is to expand the 'playing field' to other museums in the city.
Type of game and technical characteristics: Augmented reality game. Adapted for mobiles and tablets. Available on Google Play and in the App Store.
Release date: 2016.
Developer: Padaone Games.
Who is it for? Families with children from 8 years and older.
Objectives: To improve the experience of visitors to the museum.
Solution: The game presents very varied mechanics. It is a gymkhana formed by a series of funny mini-games, virtual content in 360º and tracks to advance in the adventure. The characters that will guide the player on their tour of the Galdianas Islands is the captain of Enigma, Anne Jack, and the parrot Mr Mendoza. In addition, other characters from the Lázaro collection help the child in their search for the treasure. In order to advance through the adventure tests it is necessary to find certain works of the collection, point at them with a mobile phone (hereinafter the pirate spyglass) and in this way access the content associated with that track. In addition, the game takes players into two 'magic rooms', virtual 3D scenes that the adventurer can travel through in 360º via the gyroscope of the pirate spyglass. Once the works are located, the player must overcome a series of varied mini-games: find the differences, capture an elusive pixie that hides in the works in the rooms, find the correct order of the drawers of a desk and answer questions about the works. As the player advances in their adventure, they collect a series of virtual objects that will help them to obtain the final treasure: the great sword of the Count of Tendilla. Once the main adventure (lasting about an hour) is over, the player will have the option to continue their search through the museum and find the 12 mysterious secrets of the Galdianas Islands. At the museum ticket office, parents can acquire a map of the islands with 'privileged' information to guide their children in the adventure.
Results: According to Google Play data, the app has had more than 100 installations.

Case Study 11: 'La Rioja Week' in Atrapalo

Figure 9.11 Atrapalo (please use QR code to access the image - not included due to permission issues)

Table 9.13 La Rioja Week in Atrapalo

Name, title and brief description: La Rioja Week is an advertising game through social networks. The action, integrated into the tourist plan of the region, consisted of the promotion of hotels, restaurants, activities and trips to La Rioja. It is the result of collaboration between the Atrapalo travel and leisure portal and the tourism council of the Autonomous Community of La Rioja (Salvadó, 2007).
Type of game and technical characteristics: Web microsite with information about the region. Facebook game that consisted in matching images of La Rioja.
Who is it for? Men and women aged between 25 and 50 years old.
Release date: Game presented in 2013. The agreement between Atrapalo and La Rioja Tourism Council was repeated in 2014.
Developer: Atrapalo.com.
Objectives: Promote La Rioja as a tourist destination. Increase sales of tour packages.
Solution: Online campaign boosting display actions that directed the audience to an exclusive microsite. Next, special actions with video were launched in the main home site of Atrapalo. The action was complemented with the sending of a newsletter to the million subscribers to the travel portal. Through social networks a game was launched with the theme of the promotion in which the players discovered the particularities of La Rioja. The game consisted of matching images of La Rioja, which were displayed on the Facebook page for a few moments, to get vouchers of €50.
Results: During 2013, the promotional action generated an impact on sales. The online game impressed 2000 players with an average of 12.5 games per player. It added a total of 17,000 minutes of play and 284 hours of exposure to the content of Atrápalo (Santamaría, 2014).

Once the cases had been analysed, they were classified according to the type of game dynamics and the technology and multimedia resources used (Table 9.14).

In most games analysed, geolocation platforms are used (Table 9.15). Normally, these are tours of cities or museums as a way to discover their main attractions. Sometimes the routes were complemented with the possibility of registering the visit, completing a test or answering a series of questions. There were also applications that were treasure hunts (e.g. jewels) that allowed unlocking of screens or to get some kind of reward. In the tourist world there has been a decline in contests based on social media platforms, since they tend to have little importance owing to their short duration. This indicates a downward trend in a type of action that were once widely used.

Most of the analysed examples work on mobile applications. Only in three of them were accessed through the web. All used multimedia elements (images, sounds, videos and animations). Some games complement the experience with a 360° panoramic view. Progressively, the use of augmented reality is expanding, for example, as in applications such as Amazing City Castellón, Borgia Territory, Enigma Galdiano or Aste Nagusia Guide. The Imageen Tarraco application is the only one that combined augmented reality with virtual reality. To promote virality, most games allowed sharing experiences and achievements on social networks.

Discussion and Conclusion

The review of the literature on videogames and advertising, and the study of relevant cases, allows us to draw conclusions about the phenomenon of advergaming and its situation in the tourism industry in Spain.

The saturation of the media means that the communication actions of advertisers and brands coexist in an ecosystem full of messages. The consumer exercises a more active role than before in becoming actively engaged with brands through social media and other platforms. Brands explore new ways of contact that mix entertainment and persuasion.

In this environment, videogames and tourism apps have become a place of leisure and entertainment. It is about spaces of interaction where, with quite assiduity, fan communities emerge with share common interests, values and passions.

For Ana Moles, from the Gameloft production company (personal communication, 27 April 2018), the gamified formats allow users to interact with the brand in a fun based way. Normally, in these environments users are receptive and are used to coexisting with advertising messages, obtaining a positive brand memory, as it helps them to achieve their goals. According to Moles (personal communication, 27 April 2018) 'if the game is well designed, the brand has the opportunity to show all the qualities, characteristics and advantages of its product'. The value of the brand–consumer relationship is usually measured by the degree of the engagement.

Table 9.14 Summary of the dynamics of the games

Name of the game	Dynamics of the game						
	Treasure hunt	**Quiz**	**Contest**	**Tests/ gymkhana**	**Check-in**	**Visual search**	**Geolocation route**
Amazing City Castellón	X	X					X
Imageen Tarraco						X	X
Haro in Game	X			X			X
Marco Topo in the museum		X					
The compass of Ili Palmir				X	X		X
Borgia Territory	X				X	X	X
Aste Nagusia Guide	X	X			X	X	X
Teruel Go				X	X	X	X
Province of Seville			X				
Enigma Galdiano	X			X		X	X
La Rioja Week in Atrapalo			X	X			

Table 9.15 Summary of results/technology and multimedia resources

Name of the game	Technology and multimedia resources						
	Mobile app	**Web**	**Multimedia**	**Panoramic views**	**Augmented reality**	**Virtual reality**	**Social sharing**
Amazing City Castellón	X		X	X	X		
Imageen Tarraco	X		X	X	X	X	
Haro in Game	X		X				X
Marco Topo in the museum		X	X				X
The compass of Ili Palmir	X		X				X
Borgia Territory	X		X	X	X		X
Aste Nagusia Guide	X		X	X	X		X
Teruel Go	X		X				
Province of Seville		X	X				X
Enigma Galdiano	X		X		X	X	
La Rioja Week in Atrapalo		X	X				X

From the academic world there is also optimism regarding the benefit of this type of format. Most studies analysed emphasized the validity of gamification strategies (Corrêaa & Kitanoa, 2015; Negruşa *et al.*, 2015; Sigala, 2015). The benefits range from the increase in brand awareness, the possibility of attracting users, generating a community and the creation of databases.

However, faced with the optimism of professionals and academics, there are difficulties in the development of advertising gamification. A great asymmetry of the videogame industry exists in Spain. The Spanish market has a great mismatch between expectations and real investment. In fact, there are no official data showing investment made in advergaming actions. Professor Ana Sebastian *et al.* (2016: 276–277) explains that 'there is a fear and lack of confidence on the part of the advertisers'.

Recently, we have observed an incipient development of the industry. The commercial network is still small and isolated initiatives still prevail; however, there are already companies that, after a successful experience in a territory, export it to another place. For example, the company Play & Go Experience mixes gamification, geolocation and augmented reality in its developments. After the success with its application on the Fallas of Valencia 2017, they have exported their model to other events such as the Tomatina de Buñol or the Semana Grande de Bilbao.

In most mobile applications, augmented reality is imposed as a resource that allows integrating elements of the landscape with extra information that complements the visit. In Spain, some museums such as the National Archaeological Museum (MAN) have launched interesting developments using virtual reality; however, in most cases they do not usually integrate elements of gamification.

The company Imageen combines augmented reality with virtual reality. They propose a new way of doing tourism by recreating virtually ancient Roman cities. With the help of a personality of the time, tourists can easily understand what has changed, what has survived and what has disappeared. This all occurs by entering the past through a mobile phone or, better yet, with virtual reality glasses. This project has been very successful and, since the first experience in Tarraco, applications in cities such as Cartagena, Mérida or Madrid have been developed.

In the review of the cases we find several applications that enrich the touristic visit. For example, in Amazing City Castellón tourists can live through an adventure connected to the history and gastronomy of a place in the province. The application 'Territorio Borgia' avoids the seasonality of tourism by offering a cultural product of great quality. In Teruel Go, the visitor travels round the city with the challenge of searching for and registering a series of monuments in the city by taking pictures of them.

We also find examples of tools that try to capture the attention of the younger audience. A challenge for the industry, since this section of the population needs extra motivation to visit a monument or a museum.

Thus, in the compass of Ili Palmir, a child travels through the city of Elche learning about its heritage. In the case of Marco Topo, while the parents visit the museum, the children try to solve challenges posed by the game that appears on the screen.

In summary, the use of advergaming and gamification in tourism is expanding. In recent years, the number of mobile applications that help to enrich the tourist experience through the game has grown. In the short term, technological improvements and cheaper virtual reality systems will allow widespread access to immersive experiences (Caerols *et al.*, 2020). However, analysing the download figures of some of the applications in the case studies, we note that there is still a long way to go. Shortage is noted especially in the capture of traffic to these platforms, the generation of communities and the creation and use of databases.

Most of the cases studied correspond to museums and territory brands; however, the initiatives of traditional agents in the Spanish tourism sector as providers of services (hotels, airlines, railway companies, bus companies, etc.) are lacking. For these companies, it is worth exploring gamification as a tool to help attract new audiences and build loyalty among attendees. The initiative of the Hotel Hovima la Pinta of Gran Canaria, which placed a virtual aquarium in the hall of its facilities (with more than 30 species), should not be an isolated case, but a common practice of technology integration, communication and tourism resources.

References

Anuario de la Industria del Videojuego (2016) Retrieved June 22, 2019. Available at: http://www.aevi.org.es/web/wp-content/uploads/2017/06/ANUARIO_AEVI_2016.pdf (accessed on April).

Bekerian, D.A. and Baddeley, A.D. (1980) Saturation advertising and the repetition effect. *Journal of Memory and Language* 19 (1), 17.

Bellman, S., Kemp, A., Haddad, H. and Varan, D. (2014) The effectiveness of advergames compared to television commercials and interactive commercials featuring advergames. *Computers in Human Behavior* 32, 276–283.

Buckner, K., Fang, H. and Qiao, S. (2002) Advergaming: A new genre in internet advertising. *SoCbytes Journal* 2 (1).

Buhalis, D. and Foerste, M. (2015) SoCoMo marketing for travel and tourism: Empowering co-creation of value. *Journal of Destination Marketing & Management* 4 (3), 151–161.

Buhalis, D. and Law, R. (2008) Progress in information technology and tourism management: 20 years on and 10 years after the internet – The state of eTourism research. *Tourism Management* 29 (4), 609–623.

Caerols, R., Sidorenko, P. and Garrido, P. (2020) Hacia un modelo de narrativa en periodismo inmersivo. *Revista Latina* 75, 341–365.

Campbell, C. and Marks, L.J. (2015) Good native advertising isn't a secret. *Business Horizons* 58 (6), 599–606.

Cañete, L. Martinez, I.J. and Aguado, J.M. (2013) Hacia una tipologia de advergaming y formatos afines. I Congreso Internacional de Comunicacion y Sociedad Digital, Universidad Internacional de La Rioja.

Carlson, M. (2015) When news sites go native: Redefining the advertising–editorial divide in response to native advertising. *Journalism* 16 (7), 849–865.

Castelló, A. (2012) La saturacion publicitaria en los nuevos entornos digitales: ¿una cuestion etica? In *La Etica de la Comunicacion a Comienzo del siglo XXI: I Congreso Internacional de Ética de la Comunicacion, Libro de Actas* (pp. 1128–1138). Edufora.

Çeltek, E. (2010) Mobile advergames in tourism marketing. *Journal of Vacation Marketing* 16 (4), 267–281.

Chaney, I.M., Lin, K.H. and Chaney, J. (2004) The effect of billboards within the gaming environment. *Journal of Interactive Advertising* 5 (1), 37–45.

Chen, J. and Ringel, M. (2001) Can advergaming be the future of interactive advertising. *Fast Forward*. Available at: http://www.kpe.com.

COMSCORE (2018) Informe 2018: La industria de viajes en Europa. Available at: www.comscore.com/esl/eu-travel-report.

Corrêaa, C. and Kitanoa, C. (2015) Gamification in tourism: Analysis of Brazil Quest Game. *Proceedings of ENTER* 2015 (February).

Couldry, N. and Turow, J. (2014) Advertising, big data and the clearance of the public realm: Marketers' new approaches to the content subsidy. *International Journal of Communication* 8, 1710–1726.

Covaleski, R. (2010) *Publicidade Híbrida*. Maxi Editora.

Daugherty, T., Eastin, M. S. and Bright, L. (2008) Exploring consumer motivations for creating user-generated content. *Journal of Interactive Advertising* 8 (2), 16–25.

De Mooij, M. (2018) *Global Marketing and Advertising: Understanding Cultural Paradoxes*. SAGE Publications.

Del Moral, M.E. (2014) Advergames & edutainment: Fórmulas creativas para aprender jugando. In F. Revuelta, M.R. Fernández, M.I. Pedrera and J. Valverde (eds) *Ponencia Inaugural del II Congreso Internacional de Videojuegos y Educación*, 1–3.

Del Pino, C. and Galan, E. (2010) Internet y los nuevos consumidores. El nuevo modelo publicitario. *Telos* 82, 55–63.

DEV (2014) Libro Blanco del Desarrollo Español de Videojuegos. Available at: http://www.dev.org.es/en/publicaciones/todas-las-publicaciones.

Dobrow, L. (2004) How to play games and influence people: Advergaming emerges as new ad option. *Media Daily News* 15.

Fletcher, R. and Nielsen, R.K. (2017) Are news audiences increasingly fragmented? A cross-national comparative analysis of cross-platform news audience fragmentation and duplication. *Journal of Communication* 67 (4), 476–498.

González, C. and Francés, M.T. (2009) Advergaming aplicado a las estrategias comunicativas dirigidas al target infantil. *Actas del Congreso Brand Trends*. Valencia: Universidad Cardenal Herrera-CEU.

Groh, F. (2012) Gamification: State of the art definition and utilization. *Institute of Media Informatics Ulm University* 39, 31.

Gummesson, E. (2007) Case study research and network theory: Birds of a feather. *Qualitative Research in Organizations and Management: An International Journal* 2 (3), 226–248.

Gurău, C. (2008) The influence of advergames on players' behaviour: An experimental study. *Electronic Markets* 18 (2), 106–116.

Haro, I. (2002) Advergaming. El juego de las marcas. *Estrategias de Comunicación y Marketing* 110, 60–62.

INFOADEX (2017) Resumen del Estudio Infoadex sobre Inversion Publicitaria. Available at: http://www.infoadex.es/home/wp-content/uploads/2018/02/Estudio-InfoAdex-2018.pdf.

Khan, M.L. (2017) Social media engagement: What motivates user participation and consumption on YouTube? *Computers in Human Behavior* 66, 236–247.

Klein, J.F., Falk, T., Esch, F.R. and Gloukhovtsev, A. (2016) Linking pop-up brand stores to brand experience and word of mouth: The case of luxury retail. *Journal of Business Research* 69 (12), 5761–5767.

Klophaus, R. (2005) Frequent flyer programs for European low-cost airlines: Prospects, risks and implementation guidelines. *Journal of Air Transport Management* 11 (5), 348–353.

Lambert, C. and Lee, A. (2018) Videotisements: An emerging advertising medium targeting children. Global Marketing Conference, July 2018, Tokyo, 135–140.

Law, R., Buhalis, D. and Cobanoglu, C. (2014) Progress on information and communication technologies in hospitality and tourism. *International Journal of Contemporary Hospitality Management* 26 (5), 727–750.

Lee, H. and Cho, C.H. (2017) An application of brand personality to advergames: The effect of company attributes on advergame personality. *Computers in Human Behavior* 69, 235–245.

Lee, J. and Hong, I.B. (2016) Predicting positive user responses to social media advertising: The roles of emotional appeal, informativeness, and creativity. International *Journal of Information Management* 36 (3), 360–373.

Madinaveitia, E. (2010) La publicidad en medios interactivos: En busca de nuevas estrategias. *Telos: Revista de Pensamiento Sobre Tecnología y Sociedad* 82, 43–54.

Marolf, G. (2007) *Advergaming and In-Game Advertising: An Approach to the Next Generation of Advertising*. Vdm Verlag.

Martí-Parreño, J. (2008) Videojuegos, marcas y consumidores. *MK Marketing+ Ventas*.

Martí-Parreño, J., Ruiz-Mafé, C. and Scribner, L.L. (2015) Acceptance of branded video games (advergames): A cross-cultural study Spain-USA. In *Marketing and Consumer Behavior: Concepts, Methodologies, Tools, and Applications* (pp. 851–871). IGI Global.

Martí-Parreño, J., Bermejo-Berros, J. and Aldás-Manzano, J. (2017) Product placement in video games: The effect of brand familiarity and repetition on consumers' memory. *Journal of Interactive Marketing* 38, 55–63.

Matteo, S. and Dal Zotto, C. (2015) Native advertising, or how to stretch editorial to sponsored content within a transmedia branding era. In G. Siegert, K. Förster, S.M. Chan-Olmsted and M. Ots (eds) *Handbook of Media Branding* (pp. 169–185). Cham: Springer.

McFall, L. (2004) The language of the walls: Putting promotional saturation in historical context. *Consumption Markets & Culture* 7 (2), 107–128.

Méndiz, A. (2010) Advergaming: concepto, tipología, estrategia y evolucion historica. *Icono* 14, 8 (1), 3.

Moriarty, S., Mitchell, N.D., Wells, W.D., Crawford, R., Brennan, L. and Spence-Stone, R. (2014) *Advertising: Principles and Practice*. Australia: Pearson.

Negruşa, A.L., Toader, V., Sofică, A., Tutunea, M.F. and Rus, R.V. (2015) Exploring gamification techniques and applications for sustainable tourism. *Sustainability* 7 (8), 11160–11189.

Nelson, M.R. (2005) Exploring consumer response to 'advergaming'. *Online Consumer Psychology: Understanding and Influencing Consumer Behavior in the Virtual World*, 156–182.

ONTSI (2018) Las TICs en los hogares españoles. Estudio de demanda y uso de Servicios de Telecomunicaciones y Sociedad de la Informacion. Available at: https://goo.gl/UzPGDx (accessed 12 May 2018).

Purswani, G. (2010) Advergames, their use and potential regulation. *Asia Pacific Public Relations Journal* 11 (7), 57–63.

PWC (2017) Entertainment and media outlook 2017–2021 España. https://www.pwc.es/es/entretenimiento-medios/entertainment-media-outlook-2017-2021-espana.html.

Reinares, P. and Reinares, E. (2007) Analysis of the contents of the advertising saturation on television. *Esic Market* 126, 23–94.

Rumbo, J.D. (2002) Consumer resistance in a world of advertising clutter: The case of Adbusters. *Psychology & Marketing* 19 (2), 127–148.

Salvadó, A. (2007) La Rioja Turismo y Atrapalo firman un acuerdo de colaboración. Available at: http://blogs.atrapalo.com/prensa/2013/04/30/la-rioja-turismo-y-atrapalo-firman-una-acuerdo-de-colaboracion-para-promocionar-la-rioja/#.XDIlzlxKi01 (accessed 3 January 2019).

Sánchez, F. (2006) Advergaming: La fuerza del entretenimiento. *IPMARK: Información de Publicidad y Marketing: Guía de Servicios y Proveedores* 662, 42–42.

Santamaría, C. (2014) Publicidad y videojuegos: el advergaming en el entorno de los nuevos medios sociales (undergraduate thesis). Universitat Jaume I, Castellón, España. Retrieved http://hdl.handle.net/10234/113679

Santos, E., Gonzalo, R. and Gisbert, F. (2007) Advergames: Overview. *Information, Technologies & Knowledge*, 1, 203–208.

Sebastián, A. and Carcelén, S. (2014) La importancia del advergaming como herramienta publicitaria. In T. Pintado and T. Sanchez (eds) *Nuevas Tendencias en Comunicacion Estrategia* (pp. 247–286). Madrid: ESIC.

Sebastian, A., Nuñez, M. and Muñoz, D. (2016) New business models for advertisers: The video games sector in Spain. Advergaming vs Ingame Advertising. *Revista ICONO 14 Revista científica de Comunicación y Tecnologías emergentes* 14 (2), 256–279.

Selva, D. (2009) El videojuego como herramienta de comunicacion publicitaria: Una aproximación al concepto de advergaming. *Comunicacion: Revista Internacional de Comunicacion Audiovisual, Publicidad y Estudios Culturales* 7, 141–166.

Shan, L.H., Chin, T.A., Sulaiman, Z. and Muharam, F.M. (2016) Effective mobile advertising on mobile devices. *Journal of Global Business and Social Entrepreneurship (GBSE)* 2 (4), 164–177.

Sharma, M. (2014) Advergaming – The novel instrument in the advertising. *Procedia Economics and Finance* 11, 247–254.

Shih, C.C., Lin, T.M. and Luarn, P. (2014) Fan-centric social media: The Xiaomi phenomenon in China. *Business Horizons* 57 (3), 349–358.

Sigala, M. (2015) Gamification for crowdsourcing marketing practices: Applications and benefits in tourism. In F. Garrigos-Simon, I. Gil-Pechuán and Estelles-Miquel (eds) *Advances in Crowdsourcing* (pp. 129–145). Cham: Springer.

Svahn, M. (2005) Future-proofing advergaming: A systematisation for the media buyer. *Proceedings of the second Australasian Conference on Interactive Entertainment* (pp. 187–191). Sydney: Creativity & Cognition Studios Press.

Toib, J. (2018) The best native advertising examples of 2018. Available at: https://giveitanudge.com/best-native-advertising-examples-of-2018/ (accessed 17 December 2018).

Van Reijmersdal, E.A., Lammers, N., Rozendaal, E. and Buijzen, M. (2015) Disclosing the persuasive nature of advergames: Moderation effects of mood on brand responses via persuasion knowledge. *International Journal of Advertising* 34 (1), 70–84.

Vashisht, D. and Royne, M.B. (2016) Advergame speed influence and brand recall: The moderating effects of brand placement strength and gamers' persuasion knowledge. *Computers in Human Behavior* 63, 162–169.

Viñarás, M., Cabezuelo, F. and Herranz, J.M. (2015) Corporate philosophy and brand values as axes of the new paradigm in communication. *Prisma Social* 14, 379.

Weber, J. (2014) Augmented reality gaming: A new paradigm for tourist experience. *Information and Communication Technologies in Tourism* 57–67.

Xu, F., Weber, J. and Buhalis, D. (2013) Gamification in tourism. In Z. Xiang and I. Tussyadiah (eds) *Information and Communication Technologies in Tourism 2014* (pp. 525–537). Cham: Springer.

Xu, F., Buhalis, D. and Weber, J. (2017) Serious games and the gamification of tourism. *Tourism Management* 60, 244–256.

Yang, M., Roskos-Ewoldsen, D.R., Dinu, L. and Arpan, L.M. (2006) The effectiveness of 'in-game' advertising: Comparing college students' explicit and implicit memory for brand names. *Journal of Advertising* 35 (4), 143–152.

Yılmaz, H. and Coşkun, İ.O. (2016) New toy of marketing communication in tourism: Gamification. In E. Sezgin (ed.) *e-Consumers in the Era of New Tourism* (pp. 53–71). Singapore: Springer.

10 Pokémon GO: Serious Leisure and Game-Playing Tourists

Russell B. Williams

Introduction

Pokémon GO was the first important location-based gaming experience for millions of people across the globe (Colley *et al.*, 2017). Pokémon GO is the most successful location-based augmented reality (AR) game to date. Released globally in July 2016, it was downloaded 550 million times in the first 80 days after release (Newzoo, 2016) and had been downloaded more than 750 million times by June of the following year (Smith, 2017). In the first week of release the daily average user (DAU) number was 28.5 million (Frommer, 2017). In April of 2017 there were still 5 million unique users on Pokémon GO every day, a number in the top five of mobile games globally (Smith, 2017). There are fewer people playing than at launch, but the people who are playing appear to be more involved in the game (Wandera, 2017).

Pokémon GO has had an impact on how people interact with the world and each other. It is also deeply connected with travel and tourism. Collecting Pokémon is the primary way players progress in the game (Colley *et al.*, 2017). To do that, a player must walk around and explore the physical world. Players must travel in their neighbourhood, their city, their country and the world. Pokémon are seen on a player's mobile device as virtual creatures at locations specified by the GPS in the phone. These locations include important and obscure attractions of interest in cities, towns and rural areas. Pokémon attract Pokémon *trainers*, players who also hang-out at *PokeStops* and *Gyms* that are located at places of interest in the real world. Trainers go to PokeStops to pick up Pokémon *Eggs*, which hatch into Pokémon after players walk a specified distance, *PokeBalls*, used to capture Pokémon, and other useful items for game play. PokeStops and Gyms are found at landmarks, in parks, near historical sites and places where a gathering of people is of financial or social interest.

The game, with its relatively intensive travel-oriented play and the continuing interest of a significant number of users, continues to be of academic interest. In this chapter, we discuss research findings around Pokémon GO and the thought that provides the key to gamified experiences in tourism. Pokémon GO is touted by some as the model for all future gamification efforts in tourism. We discuss how the success of Pokémon GO is nearly impossible to replicate without the connection to an enduring transmedia property.

The aim of this chapter is to consider Pokémon GO as a form of serious leisure (Stebbins, 1992) and the connection of this leisure to travel and tourism. The findings of a survey of 442 Pokémon GO players from different parts of the world demonstrates the connection between Pokémon GO play, the pursuit of serious leisure, domestic and international tourism. This study supports the positive findings from previous work on Pokémon GO (e.g. Aluri, 2017; Bonus *et al.*, 2018; Slak-Valek & Williams, 2016; Williams & Slak-Valek, 2019) but goes beyond that research considering the role of game performance as an indicator that Pokémon GO play is a form of serious leisure.

Literature Review

Tourism, games and augmented reality

Augmented Reality (AR) is an experience where a real environment is enhanced by the addition of layers of virtual elements that can include information and images (Hedley *et al.*, 2002). These virtual elements then become a part of the mediated reality presented on the device. Adding the location-based potentialities, including GPS, and increased bandwidth of modern mobile devices, AR has become a very attractive medium for many different applications (Buhalis *et al.*, 2019). Location-aware AR can display contextualised information about a place including current, historical or cultural information and anything that is not directly in the observer's field of view (Jaramillo *et al.*, 2010). This augmentation can be factual or imagined and include text, graphics, photos, sound and videos. AR applications have been developed for navigation, cultural heritage, entertainment and education. Real-time, location-sensitive tour guides, replacing human and paper-based guides, are one application that has been proposed, designed and tested (e.g. Garau, 2014; Pedrana, 2014; Yu-Lien *et al.*, 2015). These guides can portray the current state of an attraction and provide insights into its past or developments for the future. The full potential for AR in tourism is still being explored and developed (Yovcheva *et al.*, 2012).

Yovcheva *et al.* (2014) found that tourists had a positive attitude toward AR, the ability to add contextualised information to the environment and the ease of use. However, with prolonged use these respondents gave increased negative comments and expressed dissatisfaction with some content and interface elements of the experimental app. In another piece of research, individuals using AR for guidance learned more than participants using other, more traditional, means of guidance and the technology helped them 'construct a memory of the place' (Yu-Lien *et al.*, 2015: 174). Chung *et al.* (2015) found that personal factors and visual appeal had an impact on the perceived usefulness of AR for heritage tourists. Jung *et al.* (2015) found that content quality, system quality and the quality of personalised service in a theme-park-based AR app positively influenced visitors' sense of satisfaction and that satisfaction led to positive word-of-mouth for the park. They also found that all users valued personalised service quality.

Kourouthanassis *et al.* (2015) found that the level of personal innovativeness was important in initial user engagement with an AR app for the Olympia temple in Greece and that consistency in content and usefulness were important in continued use. They emphasised the importance of user interface and personalisation. They concluded that the AR system enhanced motivation, engagement and the overall tourist experience, and that the system could be improved through the introduction of game elements.

Games can also be a significant part of tourism experiences. Mesáro *et al.* (2016) argue that gamification of destinations and attractions can be a way to attract visitors and increase interaction between users and the experiences they encounter. Mortara *et al.* (2014) found that gamified experiences in heritage tourism take many forms, including trivia, puzzles, mini-games, adventures, role playing and simulations. Some of these are for desktop and laptop devices while others are for mobile devices. The context of the games included museums, exhibitions, real-world sites and attractions. Xu *et al.* (2016) also emphasised the potential of gaming for tourism, whether it is social games as attractors or location-based games as engagers. They described AR as a popular way to improve immersive experiences in location-based tourist games. The gaming element appears to intensify the feeling of immersion in the AR experience.

McCall and Braun (2008) and McCall *et al.* (2011) reported on a city-wide mobile AR gaming experience in Cologne, Germany. They found that once users disengaged from the hybrid game world to navigate the city they had less interest in returning to the game. Makkonen and Hokkanen (2013) found that a mobile game in a Finnish museum had very little impact on tourists and their decision to visit. They concluded that 'A "people will come" mindset functions poorly in tourism' (Makkonen & Hokkanen, 2013: 266) regardless of how much innovation is included in the new approach.

Pokémon GO and Travel

Pokémon GO is built on the foundation of the Pokémon universe and the 20 years of cross-media success that includes digital games, trading-card games, television series and films. Pokémon is the second most successful videogame franchise after Mario Brothers (Zsila *et al.*, 2017). Rodríguez-Serrano *et al.* (2017) found that 75% of the players of Pokémon GO they studied watched the Pokémon movies and TV series. Colley *et al.* (2017) found that 79% of their respondent players were already Pokémon fans.

The tagline is 'Gotta Catch 'Em All' and it is the core premise and apparent attraction of the franchise. Even though the initial release of Pokémon GO only contained 151 Pokémon to capture and train there are actually over 700 creatures in the Pokémon Universe. Wilson's Biophilia Hypotheses indicate that humans have an innate desire to catalogue, study and spend time with other life-forms (Balmford *et al.*, 2002). Every facet of Pokémon taps directly into this desire. Lazzaro (2009) identifies collection as an important aspect of serious fun when playing games. Assunção *et al.* (2017) identify collection as the core idea for the creation of the Pokémon Universe. Fifty-two percent of Aluri's (2017) respondents saw the ability to review captured Pokémon and game inventory enhancing the overall experience of the app and 72% enjoyed the ability to capture Pokémon and interact with the augmented world. Rodríguez-Serrano *et al.* (2017) found that the main reason for playing among their respondents was the capture of new Pokémon.

The travelling aspect of Pokémon is directly connected to collecting because a player must go to specific places to find particular pokémon. McIntosh (2016) insists that Pokémon as a franchise has always embraced adventure, travel and tourism. He says he took the time to travel his hometown and province just to find creatures playing Pokémon GO. This was also in the minds of Pokémon GO creators as John Hanke stated, 'The basic notion was there's a lot of cool history and lore and unknown secrets in your own neighborhood that you don't know about' (Parker, 2016: 41). Gretel-Kauffman (2016) pointed out that an agreement between *Niantic* (the developers and distributors of Pokémon GO), Google and the Historical Marker Database helped create PokeStops at historical markers, memorials, plaques, statues and landmarks. Players have expressed appreciation for what they have learnt in pursuit of Pokémon. Amselle (2016) discussed in length Pokémon GO's impact on parks as these spaces hold both quantity and quality of Pokémon within the game. Aluri (2017) found that 58% of respondents were curious to know more real-world information behind PokeStops they visited and concluded that unnoticed landmarks could become popular because of the game. However, Aluri (2017) also found that limited interaction with the real world was one of the top three things that players disliked about playing Pokémon GO.

An impact of Pokémon GO on travel and exploration is the changing of players established patterns of movement. In the pursuit of Pokémon, players changed how they moved through space and subsequently experienced more of the space and places around them. Huiwen *et al.* (2017) emphasised the predictability of human movement between a number of fixed locations and the fact that Pokémon GO has changed these patterns. Colley *et al.* (2017) found that 60% of their respondents visited at least one new place while playing Pokémon GO and 9% visited an entirely new town or city for play. Williams and Slak-Valek (2019) found that Pokémon GO players travel more domestically and internationally, plan their lives and days more carefully to enable play. They spend less time at home in front of a screen playing console games.

Assunção *et al.* (2017) found respondents willing to play because of the nostalgia associated with the original group of 151 Pokémon creatures. Zsila *et al.* (2017) found nostalgia as the second highest motivation to play the new game and other researchers have focused on the power of Pokémon nostalgia as well (Bonus *et al.*, 2018; Loveday & Burgess, 2017; Tang, 2017; Vaterlaus *et al.*, 2018; Yang & Liu, 2017).

Serious leisure

Serious leisure is a concept first identified by Stebbins (1982) to help explain why and how people pursue particular activities outside of their non-working time. There are six fundamental elements in the pursuit of serious leisure (Moularde & Weaver, 2016; Xiangyou & Yarnal, 2010): (1) perseverance even in the face of difficulties or obstacles; (2) progressive development of skills and/or knowledge with levels of achievement over the course of a career in the endeavour; (3) the employment of significant personal effort and practice that includes learning from peers, tutors and specialised media; (4) lasting benefits in the form of self-actualisation, self-esteem and social interaction;(5) a unique subculture, or ethos, that bonds participants and makes them identifiable to others through their values, behaviours and language; and (6) social identification linked with the activity and the attendant sense of belonging.

One only needs to think of 'gamer' culture to get a picture of serious leisure. Players of Pokémon GO demonstrate these attributes as do people who pursue sports, seemingly complex hobbies and niche forms of tourism in their leisure time. Not all participants in serious leisure are the same and their levels of seriousness and specialisation exist across a spectrum (Scott, 2012) from casual (Stebbins, 1997) to serious. Here, we identify the mid-range on that spectrum as transitional (Figure 10.1).

Figure 10.1 Serious leisure spectrum

Serious leisure has been studied for many different groups of people. Curtin (2010) found the principles in operation with wildlife watchers of which Pokémon GO is an AR version. Suni (2017) found evidence of serious leisure and connected it to touristic activities of Finnish hunters, possibly another activity analogous to Pokémon GO in the non-augmented world. Surfers and their travel choices have also demonstrated the explanatory power of serious leisure (Cristina-Portugal *et al.*, 2017) as have mountain bikers (Moularde & Weaver, 2016). Online gamers have also been found operating on the serious leisure spectrum (Wu *et al.*, 2013). Though not connected with the serious leisure perspective, Pokémon GO players have been described according to the six principles (Lalot *et al.*, 2017). Green and Jones (2005: 177) could have been describing Pokémon GO players when they described serious sports tourists having 'knowledge of places, events and attractions, knowledge of subcultural stories and lore, experience in the ritual and ceremony of the subculture'. The activities of Pokémon GO players can clearly be understood in these terms.

Pokémon GO play is one variety of serious leisure and the touristic activities associated with it are indicative of where a player falls on the spectrum of seriousness and specialisation. Based on the theory, serious Pokémon GO players should be significantly different than casual players in terms of their travel activities and their awareness of the places they have encountered. Based on the literature and the nature of success in the game serious players should report playing in more foreign countries, playing in more different cities, and more travel overall as a result of Pokémon GO play. Because of their identification as serious players and the connection of Pokémon GO to landmarks and attractions, they should also report more visits to tourist attractions, greater knowledge of local tourist attractions, greater knowledge of national tourist attractions, and greater knowledge of historical and cultural sites. Research and theory also indicate that serious Pokémon GO players should also report increased levels of happiness on the basis of play. This is a basic construct in serious leisure (Stebbins, 1992). A variance that has no background in the literature is unplanned travel. Individual people travel in very different ways and it is interesting to look at how the serious play of Pokémon GO impacts this most fundamental approach to tourism.

While the relationships between serious and casual players is clear from the literature, it is less clear how transitional players differ from these two groups on the same measures. People engaged in transitional leisure have not been previously identified and analysed, so it is uncertain where they express significant differences from the other two groups. The differences and similarities are discussed here to lay the groundwork for further examination of transitional leisure in a wide variety of settings.

There are then eight hypotheses and two research questions reported in this chapter:

(1) Hypothesis 1: Serious players will report playing in more foreign countries than casual players.
(2) Hypothesis 2: Serious players will report playing in more different cities than casual players.
(3) Hypothesis 3: Serious players will report more travel for play than casual players.
(4) Hypothesis 4: Serious players will report more visits to tourist attractions than casual players.
(5) Hypothesis 5: Serious players will report greater knowledge of local tourist attractions than casual players.
(6) Hypothesis 6: Serious players will report greater knowledge of national tourist attractions than casual players.
(7) Hypothesis 7: Serious players will report greater knowledge of historical and cultural sites than casual players.
(8) Hypothesis 8: Serious players will report higher levels of happiness on the basis of play than casual players.
(9) Research Question 1: What is the impact of serious play on the report of unplanned travel?
(10) Research Question 2: What is the relationship between those who play in the transition area between serious and casual leisure and those who play seriously and casually.

These hypotheses and research questions all connect with the assumption that touristic activities associated with Pokémon GO are indicators of serious leisure and not as a game that serves as the *holy grail* for promotion and engagement in many different forms tourism. If these hypotheses are supported then it is possible to conclude that Pokémon GO is serious leisure and a specialised form of tourism practised by people who are playing the game. These are people who can be attracted and engaged but on the basis of the game and not a more traditional form or category of tourism.

Methodology

Questionnaire development

This research is explorative as it examines Pokémon GO from the serious leisure perspective for the first time. To test the hypotheses and research questions presented here, and others reported elsewhere, a self-administered questionnaire was developed to measure changes in travel attitudes and experiences on a five-point Likert scale, where 1 meant 'strongly disagree' and 5 meant 'strongly agree'. Scale items included 'I travel more since playing Pokémon GO', 'I know more tourism attractions in my own city since playing' and 'I know more about historical and cultural sites since playing'. Simple self-report items included 'In how many countries have you played Pokémon GO?', 'In how many cities have you played Pokémon GO?' and 'How many Pokémon do you have in your *Pokedex*?' These items were developed through discussions with tourism and game studies colleagues as well as a focus group of players and then confirmed through pilot testing. Happiness items were based on Nawijin's (2010) holiday happiness curve.

A focus group discussion with seven experienced Pokémon GO players from three different countries (three from Japan, three from Slovenia and one American) was organised before finalising the questionnaire. After improving the questionnaire and taking into consideration the focus group results, it was posted online and sent to 15 Pokémon GO players with a request to test it. Pre-testers did not report any technical problems or unclear wording and the survey was activated online.

Data collection, sample and reliability

Data collection began on 18 September 2016, which is only two months after the game launch. Domahidi *et al.* (2014) found that mobile game players are active social media users. Thus, an online survey was created on www.1ka.si and posted on social media in order to reach respondents from around the world. Links to the survey were posted on the social media accounts (Facebook, Instagram and Twitter) of the author and colleagues, asking friends and followers to repost and share. A snowball (viral) method was planned.

The questionnaire did go viral, but did not reach specifically Pokémon GO players: 919 people clicked on the introductory page on September 18, but only 39 completed the survey. The introductory page was an invitation to Pokémon GO players to complete the survey. Consequently, the link to the survey was posted on Pokémon GO Facebook-groups of different countries/regions/cities specifically created for the players to exchange experiences and comments about the game, places and Pokémon creatures. In this way the target group of Pokémon GO players was reached.

The survey was active until 20 December 2016. There were 647 respondents over the age of 18 who moved past the first page of the survey and 442 (68%) completed enough of the survey to be used in this analysis. These respondents came from 34 different countries.

A reliability analysis was performed for the seven items measuring travel changes. A Cronbach's alpha value of 0.866 showed that this portion of the questionnaire achieved internal consistency across respondents of various nationalities. There were also questions about the number of cities and countries where the respondent have played the game. Seriousness of play was measured on a single item asking how many

Pokémon the player had in his or her Pokedex. Success in the game is measured by the collection of Pokémon, which are recorded in the player's Pokedex. Success can only be achieved through diligent effort and a sense of seriousness about the game. This count of Pokémon is a significant indicator that a person is making the game into an avocational career.

Correlations between the number of Pokémon collected and the tourism-related variables of interest were measured using Pearson's *r*. This analysis serves as an indicator of the direction and strength of the relationship between the level of seriousness in the game and travel-related cognitions and behaviours. In order to demonstrate specific differences between casual, transitional and serious players of the game, three groups were created on the number of Pokémon in the Pokedex (M = 94.07, SD = 31.88). Serious players had a Pokedex number that was above the mean plus ½ standard deviation (>110). Casual players had a Pokedex number that was below the mean minus ½ standard deviation (<80). Transitional players had a Pokedex number that was between the mean minus ½ standard deviation and the mean plus ½ standard deviation (80–110). To test the hypotheses, one-way ANOVAs were calculated on the means for the responses to the items on the basis of the three groups. Tukey HSD post-hoc tests were then performed on one-way ANOVAs with statistical significance.

Results

Travel between countries: Hypothesis 1

Hypothesis 1 is that serious players will report playing in more foreign countries than casual players. A Pearson correlation coefficient was computed to assess the relationship between the size of a respondent's Pokémon collection and the number of countries where the respondent had played Pokémon GO (Table 10.1). There was a significant positive correlation between the two variables. This means that when a player has collected more Pokémon then they have visited more foreign countries than other players.

A one-way between-respondents ANOVA was conducted to compare serious players, transitional players and casual players and the number of countries they had played Pokémon GO. There was a significant effect of

Table 10.1 Pearson correlations between size of Pokedex and dependent variables

	Between countries	Between cities	Play travel	Attraction visits	Local knowledge	National knowledge	History culture	Happy
Number of Pokémon collected	0.133**	0.167***	0.246***	0.280***	0.233***	0.230***	0.166***	0.229***

**$P < 0.01$; ***$P < 0.001$

serious leisure levels at the $P < 0.05$ level on the number of countries visited for play [$F(2, 439) = 4.12$, $P = 0.017$]. Post hoc comparisons using the Tukey HSD test indicated that the mean score for the serious players ($M = 1.69$, $SD = 1.02$) was significantly different than the casual players ($M = 1.42$, $SD = 0.702$) and the transitional players ($M = 1.46$, $SD = 0.838$). However, transitional players did not significantly differ from casual players on this measure.

Hypothesis 1 is supported. Serious Pokémon trainers played in significantly more countries than casual trainers. Serious trainers also played in more countries than transitional trainers. Serious trainers will go beyond their corner of the world to advance within the game and the Pokémon community of players. Players pursuing Pokémon GO as a form of serious leisure are likely to visit more foreign countries to play than other players of the game. Having interesting or unique Pokémon available within a specific and location country could attract these serious players. This adds the ideas behind serious leisure to the importance of collection in Pokémon GO found by Aluri (2017) and Rodríguez-Serrano *et al.* (2017).

Travel between cities: Hypothesis 2

Hypothesis 2 states that serious players will report playing in more different cities than casual players. A Pearson correlation coefficient was computed to assess the relationship between the size of a respondent's Pokémon collection and the number of cities where the respondent had played Pokémon GO. There was a significant positive correlation between the two variables. This means that players with more Pokémon are likely to have visited a more diverse number of cities to collect them.

A one-way between-respondents ANOVA was conducted to compare the serious players, transitional players and casual players and the number of cities where they had played Pokémon GO (Table 10.2). There was a significant effect of serious leisure on the number of cities where the game had been played. Seeking to support this hypothesis post hoc comparisons (Table 10.3) indicated that the mean number of other cities for the serious players ($M = 3.04$, $SD = 1.38$) was significantly different than the casual players ($M = 2.57$, $SD = 1.09$) and the transitional players ($M = 2.61$, $SD = 1.20$). Once again, transitional players did not significantly differ from casual players on this measure.

Hypothesis 2 is supported. Serious trainers played in significantly more cities than casual trainers. Serious trainers also played in more cities than transitional trainers. According to these first two measures it appears transitional players approach the game in the same way as casual players. They are not as proactive about succeeding in the game as serious players and are less willing to travel outside of their comfort zone to collect Pokémon. Pokémon GO play as serious leisure can draw players to cities outside of their everyday experience and DMOs can take advantage of this

Table 10.2 Summary of ANOVA for Hypothesis 2

	Sum of squares	df	Mean square	F
Between groups	19.608	2	9.804	6.493
Within groups	662.890	439	1.510	
Total	682.498	441		

Table 10.3 Tukey HSD comparison for cities played

			95% Confidence interval	
Comparisons	**Mean weight difference**	**Standard error**	**Lower bound**	**Upper bound**
Serious vs transitional	0.436**	0.140	0.11	0.77
Serious vs casual	0.474**	0.150	0.12	0.83
Casual vs transitional	–0.038	0.143	–0.37	0.30

**P <0.01

opportunity to attract potential tourists. This supports the desire to collect Pokémon found by Aluri (2017) and Rodríguez-Serrano *et al.* (2017) and connects that desire to travel through the intensity of serious leisure (Moularde & Weaver, 2016). Colley *et al.* (2017) found that new places and entirely new towns and cities were visited by players involved in the game.

General travel: Hypothesis 3

Hypothesis 3 states that Pokémon GO players who pursue the game as serious leisure will report a higher level of travel than those who pursue it casually. A Pearson correlation coefficient supported this relationship between the size of a respondent's Pokémon collection and their report of increase in general travel because of Pokémon GO. There was a significant positive correlation between the two variables.

A one-way between-respondents ANOVA was conducted to compare the serious players, transitional players and casual players and test the hypothesis (Table 10.4). There was a significant effect of serious leisure levels. Post hoc comparisons showed that the mean score for the serious players (M = 3.40, SD = 1.52) and transitional players (M = 3.41, SD = 1.42) were each significantly different than the casual players (M = 2.49, SD = 1.44). The difference in serious and transitional players was negligible (Table 10.5).

Hypothesis 3 is supported. Serious trainers reported a greater increase in travel as a result of play than casual trainers. Transitional trainers also reported more travel than casual trainers. Considering this as the least

Table 10.4 Summary of ANOVA for Hypothesis 3

	Sum of squares	df	Mean square	F
Between groups	76.264	2	38.132	17.854 ***
Within groups	937.591	439	2.136	
Total	1013.855	441		

***P <0.001

Table 10.5 Tukey HSD comparison for general travel

			95% Confidence interval	
Comparisons	**Mean weight difference**	**Standard error**	**Lower bound**	**Upper bound**
Serious vs transitional	–0.015	0.166	–0.41	0.38
Serious vs casual	0.903***	0.178	0.48	1.32
Casual vs transitional	–0.918***	0.170	–0.28	-0.52

***P <0.001

specific measure of travel increase, transitional players move away from casual players and see themselves pursuing the game in a way that is directly reflective of serious play. This means that both serious and transitional players are willing to travel to play Pokémon GO and this can be used by any kind of destination or attraction to draw players domestically and internationally. McIntosh (2016) said that the core of the Pokémon franchise is travel and that is clearly evidenced within this group of respondents.

Increased visits to tourist attractions: Hypothesis 4

Hypothesis 4 states that serious players will report more visits to tourist attractions than casual players. A Pearson correlation coefficient was computed to assess the relationship between the size of Pokémon collections and the report of visits to tourist attractions. There was a significant positive correlation between the two variables: $r = 0.280$, n = 442, $P = 0.000$.

A one-way between-respondents ANOVA was conducted to compare the serious players, transitional players and casual players and their increased level of tourist attraction visits because of Pokémon Go. There was a significant effect of serious leisure levels at the $P < 0.05$ level on the increase in visits to tourist attractions [$F(2, 439) = 16.06$, $P = 0.000$]. Post hoc comparisons using the Tukey HSD test indicated that the mean score for the serious players (M = 3.55, SD = 1.36) was significantly different than the casual players (M = 2.66, SD = 1.44) but not than transitional players (M = 3.35, SD = 1.28). Casual players also significantly differed from transitional players.

Hypothesis 4 is supported. Serious trainers reported greater increases in visits to tourist attractions than casual trainers. Transitional trainers also reported greater increases in tourist attraction visits than casual trainers. This increase in tourist attraction visits is good for attraction managers, but this measure does not define whether these attractions are local for the player or distant. These could be hometown tourist attractions and this localised interpretation separates transitional players from casual players and makes transitional players appear to be moving towards the actions taken by serious players. Individuals who play the game as serious and transitional leisure do report higher levels of visits to tourist attractions because of their play. This could provide a localised group of visitors who might not otherwise be interested in their local attractions. This is consistent with the potential described by Mesáro *et al.* (2016) and Xu *et al.* (2016).

Increased knowledge of local tourist attractions: Hypothesis 5

Hypothesis 5 suggests that serious players will report a greater knowledge of local tourist attractions than casual players. A Pearson correlation coefficient supports this relationship between the size of a respondent's Pokémon collection and their increase in knowledge of local city tourist attractions because of Pokémon GO. There was a significant positive correlation between the two variables. Having more Pokémon connects with reporting more knowledge about local tourist attractions.

In order to substantiate this hypothesis a one-way between-respondents ANOVA was conducted to compare the serious players, transitional players and casual players and their increased level of knowledge of local tourist attractions because of Pokémon Go. There was a significant effect of serious leisure levels at the $P < 0.05$ level on the increase in knowledge of local tourist attractions [$F(2, 439) = 9.05$, $P = 0.000$]. The Tukey HSD test indicated that the mean score for the casual players ($M = 3.03$, $SD = 1.49$) was significantly different than the serious players ($M = 3.73$, $SD = 1.32$) and the transitional players ($M = 3.43$, $SD = 1.28$). Serious players did not have a statistically significant difference from transitional players.

Hypothesis 5 is supported. Serious trainers reported greater increased knowledge of local tourist attractions than casual trainers. Transitional trainers also reported greater increased knowledge of local tourist attractions than casual trainers. On this measure, transitional players reinforce that they are not the same as casual players and their experience of the game is closer to that of serious players. The location of Pokémon and PokeStops have direct connections to interesting locations, landmarks and tourist attractions. This increased knowledge of serious players is a side benefit of their attitude towards the game. This can be used to draw players to these locations. Once again game play has shown itself to be valuable in teaching people about the local attractions they visit. This is consistent with the potential described by Mesáro *et al.* (2016) and Xu *et al.* (2016). Aluri (2017) found that over half of the players studied wanted to know more about the places they went while playing the game. The key is to draw game players into a place to play and then it becomes possible for them to learn about the place as well as play.

Increased knowledge of national tourist attractions: Hypothesis 6

Hypothesis 6 states that serious players will report greater knowledge of national tourist attractions than casual players. A Pearson correlation was used to test the relationship between the size of a respondent's Pokémon collection and their increase in knowledge of national tourist attractions because of Pokémon GO. There was a significant positive correlation between the two variables. Larger collections of Pokémon are also significantly connected with increased knowledge about national tourist attractions.

A one-way between-respondents ANOVA was conducted to compare the serious players, transitional players and casual players and their increased level of knowledge of national tourist attractions because of Pokémon Go (Table 10.6). There was a significant effect of serious leisure levels on knowledge of national tourist attractions. The Tukey HSD test indicated that the mean score for the casual players (M = 2.72, SD = 1.39) was significantly different than the serious players (M = 3.40, SD = 1.35) and the transitional players (M = 3.18, SD = 1.28). Serious players did not differ significantly from transitional players (Table 10.7).

Hypothesis 6 is supported. Casual Pokémon trainers have lower increases and differences that are statistically significant in relation to serious trainers. There is not a statistically significant difference between this increased knowledge in transitional trainers and serious trainers, but serious players do have higher levels on all of the knowledge measures. This finding is a logical extension of the findings in Hypothesis 5. It still points to the fact that in local and national domestic terms transitional players look to be more similar to serious players than casual players. This is another factor of serious and transitional players that can be used by

Table 10.6 Summary of ANOVA for Hypothesis 6

	Sum of squares	df	Mean square	F
Between groups	32.556	2	16.278	9.356***
Within groups	763.788	439	1.740	
Total	796.344	441		

****P* <0.001

Table 10.7 Tukey HSD comparison for national attractions knowledge

Comparisons	Mean weight difference	Standard error	95% Confidence Interval	
			Lower bound	Upper bound
Serious vs transitional	0.211	0.150	–0.14	0.56
Serious vs casual	0.680***	0.161	0.30	1.06
Casual vs transitional	–0.470**	0.153	–0.83	–0.11

**P <0.01; ****P* <0.001

domestic DMOs to bring players to their attractions, attractions that might not otherwise be given much attention by the local population. This is consistent with the potential described by Mesáro *et al.* (2016) and Xu *et al.* (2016). This also connects with Aluri's (2017) finding that over half of the players studied wanted to know more about the places they went while playing the game.

Increased knowledge of historical and cultural sites: Hypothesis 7

Hypothesis 7 indicates that serious players will report greater knowledge of historical and cultural sites than casual players. A Pearson correlation was computed to assess the relationship between a respondent's collection of Pokémon and their reported increase in knowledge of historical and cultural sites since playing Pokémon GO. There was a significant positive correlation between the two variables. Though the correlation is not a strong as with local and national tourist attractions there is an indication that more serious play results in increased knowledge about historical and cultural sites.

In order to test the hypothesis after establishing this correlation a one-way ANOVA was used to compare the serious players, transitional players and casual players and their increased level of knowledge of historical and cultural sites (Table 10.8). There was a significant effect of serious leisure levels. Post hoc comparisons indicated that the mean score for the casual players (M = 2.50, SD = 1.35) was significantly different than the serious players (M = 3.04, SD = 1.35) and the transitional players (M = 2.88, SD = 1.28) (Table 10.9). Serious players did not differ significantly differed from transitional players. Pokémon Go is creating increased cultural and historical knowledge along with other types of attractions for serious and transitional players.

Hypothesis 7 is supported. Casual trainers and serious trainers have differences on this measure that are statistically significant. The differences between serious trainers and transitional trainers have substance but they are not statistically significant. Cultural and historical sites can

Table 10.8 Summary of ANOVA for Hypothesis 7

	Sum of squares	df	Mean square	F
Between groups	20.951	2	10.476	5.973**
Within groups	769.929	439	1.754	
Total	790.880	441		

**P <0.01

Table 10.9 Tukey HSD comparison for historical and cultural site knowledge

Comparisons	Mean weight difference	Standard error	95% Confidence interval Lower bound	Upper bound
Serious vs transitional	0.159	0.151	–0.20	0.51
Serious vs casual	0.543**	0.162	0.16	0.92
Casual vs transitional	–0.384*	0.154	–0.75	–0.02

*P <0.05; **P <0.01

also benefit from connection with the game and attracting these players to their locations. This is once again consistent with the potential described by Mesáro *et al.* (2016) and Xu *et al.* (2016). The desire for increased knowledge about the places visited in play was also found by Aluri (2017) in relation to Pokémon GO.

Increased happiness: Hypothesis 8

Hypothesis 8 indicates that serious players will report higher levels of happiness on the basis of play than casual players. A Pearson correlation coefficient was computed to assess the relationship between the size of a respondent's Pokémon collection and their increase in happiness because of Pokémon GO. There was a significant positive correlation between the two variables. This means there is a relationships between level of happiness and the number of Pokémon that have been collected.

A one-way between-respondents ANOVA was conducted to compare the serious players, transitional players and casual players and their increased level of happiness as a result of Pokémon GO. There was a significant effect of serious leisure levels at the $P < 0.05$ level on the increase in happiness as a result of play [$F(2, 439) = 13.31$, $P = 0.000$]. Post hoc comparisons using the Tukey HSD test indicated that the mean score for the transitional players ($M = 3.58$, $SD = 1.21$) was nearly the same as the score for the serious players ($M = 3.65$, $SD = 1.23$). Casual players ($M = 2.93$, $SD = 1.37$) were significantly lower than serious players and transitional players. Pokémon GO is related to increased happiness for the more deeply engaged players.

Hypothesis 8 is supported. There is a statistically significant difference in the feeling of happiness for serious trainers and casual trainers. Serious trainers have a greater sense of happiness since playing the game than casual trainers. Transitional trainers are also different from casual trainers and closer to serious trainers. When an attraction or destination invites game players into their world by affording the play of a game like Pokémon GO the associated level of happiness may be connected to the place and then passed along as the players recount their adventures in the place to others (Jung *et al.*, 2015).

Increased unplanned travel: Research Question 1

Research Question 1, which is explored for the first time in this study, asks whether a serious approach to play has an impact on person's report of unplanned travel. A Pearson correlation coefficient indicated a positive relationship between the size of a respondent's Pokémon collection and their increase in instances of unplanned travel.

A one-way between-respondents ANOVA was indicated more about the relationship between the serious players, transitional players and casual players and their increased level of unplanned travel because of Pokémon GO (Table 10.10). There was a significant effect of serious leisure levels. Post hoc comparisons indicated that the mean score for the transitional players ($M = 3.32$, $SD = 1.42$) was significantly different than the casual players ($M = 2.64$, $SD = 1.42$) but not the serious players ($M = 3.59$, $SD = 1.34$). Serious players also differed significantly from casual players. Pokémon GO is creating more moments of spontaneity for serious and transitional players than for casual players. Players with greater engagement in the game have greater opportunities for improvisation in their play as they decide on their next destination in the moment and not according to a specific plan.

While serious trainers have increased action, knowledge and happiness they are also more unplanned in their travel activities. The serious trainers are more willing to follow where the game leads them and perhaps they realise that this precludes advance planning of what will happen during play and travel. This is another aspect of serious leisure that can be exploited by DMOs as the highly-engaged players are more willing to follow the game rather than a personally devised itinerary. The game can be used to lead visitors into new or under-experienced areas, exhibitions and attractions that may not have the highest priority in the highly planned adventures of others. Huiwen *et al.* (2017) also found that the game was completely altering player's spatial and navigational habits, taking them outside of the comfort zone of their ritualised plans.

Table 10.10 Summary of ANOVA for Research Question 1

	Sum of squares	df	Mean square	F
Between groups	64.758	2	32.379	16.615***
Within groups	855.506	439	1.949	
Total	920.265	441		

****P* <0.001

Table 10.11 Tukey HSD comparison for unplanned travel

			95% Confidence interval	
Comparisons	**Mean weight difference**	**Standard error**	**Lower bound**	**Upper bound**
Serious vs transitional	0.266	0.159	–0.11	0.64
Serious vs casual	0.951***	0.170	0.55	1.35
Casual vs transitional	–0.685***	0.162	–1.07	–0.30

****P* <0.001

Transitional leisure and casual leisure: Research Question 2

Research Question 2 involves the introduction of transitional leisure in this study. Transitional players and casual players were found to be different. Transitional players and serious players were also different on a few measures. Those differences moved in both directions which serves to confirm the transitional nature of these players' level of interest in the game. When the serious player is ready to play in a foreign country or a different city from home then transitionals are closer to casuals. They have not made the almost professional commitment of serious leisure pursuits (Stebbins, 1982). On other measures of action, knowledge and enjoyment, transitionals are not far different than serious players. Transitionals are not casual in their play but they have not made the level of commitment that can be found with serious players. Transitional is a distinct approach on the spectrum from casual to serious. Responding to the most general item on increased travel transitionals in this study reported a slightly higher increase in travel in relation to Pokémon GO than serious players. These findings indicate that transitionals are not as locked into the game and progression as serious players, particularly in terms of travel away from home. Yet transitionals are experiencing more travel-related impacts from gameplay than casuals. Transitional Leisure is differentiated from both serious and casual leisure and not in a simple or uni-directional way. Scott (2012) established that levels of seriousness and specialisation in leisure exist across a broad spectrum and this research has established another point on the range.

Table 10.12 Summary of outcomes on eight hypotheses and two research questions

	Status	Explanation
Hypothesis 1: Serious leisure and foreign countries	Supported	The number of foreign countries visited since playing are significantly higher for those who approach Pokémon GO as serious, rather than casual, leisure.
Hypothesis 2: Serious leisure and different cities	Supported	The number of different cities visited since playing are significantly higher for those who approach Pokémon GO as serious, rather than casual, leisure.
Hypothesis 3: Serious leisure and travel for play	Supported	The amount of travel to play the game is significantly higher for those who approach Pokémon GO as serious, rather than casual, leisure.
Hypothesis 4: Serious leisure and tourist attraction visits	Supported	The number of visits to tourist attractions are significantly higher for those who approach Pokémon GO seriously rather than casually.
Hypothesis 5: Serious leisure and local attraction knowledge	Supported	Reports of local attraction knowledge since playing are significantly higher for those who approach Pokémon GO as serious, rather than casual, leisure.
Hypothesis 6: Serious leisure and national attraction knowledge	Supported	Reports of national attraction knowledge since playing are significantly higher for those who approach Pokémon GO as serious, rather than casual, leisure.
Hypothesis 7: Serious leisure and historical/ cultural site knowledge	Supported	Reports of historical and cultural site knowledge since playing are significantly higher for those who approach Pokémon GO as serious, rather than casual, leisure.
Hypothesis 8: Serious leisure and happiness levels	Supported	Happiness levels since playing are significantly higher for those who approach Pokémon GO as serious, rather than casual, leisure.
Research Question 1: Serious leisure and unplanned travel	The level of unplanned travel since playing is higher for those who approach Pokémon GO as serious, rather than casual, leisure.	
Research Question 2: Transitional leisure	Players who are identified in a position of transitional leisure are closer to casual leisure players in their travel to foreign countries and different cities but are closer to serious leisure players on every other measure.	

Discussion

There are three reasons that Pokémon GO is a guidepost for understanding and engaging gamers as tourists and is not the holy grail for the gamification of tourism experiences, marketing and engagement.

The first reason that Pokémon GO is not the holy grail for the gamification of tourism is the link between Serious Leisure and Pokémon GO play found in this study. All eight hypotheses on the differences between serious and casual play are supported. The game is still approached by some as an activity to casually pass time, but even that paradigm of play requires local travel and engagement with local places. The game can be, and is, also approached with the focus and determination of serious leisure. This is a game that has a culture and history surrounding it and has been building a fan-base for 20 years. It is more than a one-hit wonder or the shooting star of location-based augmented reality. When the trend followers stopped playing before the end of the summer of 2016 there was still a more than substantial base of players to sustain continued development and promotion, 5 million people every day. This is a significant niche that can be activated and attracted on the basis of their desire to play, advance and achieve in the Pokémon Universe. The pursuers of serious leisure are the key to this niche but the newly identified transitional players could increase the benefits of exploiting this niche as they express some of the attitudes and take some of the actions of serious players. Transitionals may even have the potential to become serious with the offer of appropriate incentives and experiences. Pokémon trainers of all varieties will come to play the specific game when destinations, attractions and enterprises offer them the game play and scoring opportunities they seek. Casuals and transitionals are less likely to play farther from their home than serious players but they are still playing the game.

The second reason that Pokémon GO is not the holy grail for the gamification of tourism experiences is the importance of nostalgia for Pokémon GO play. 'Its success is due to its perfect match, in that it allows players who had childhood fantasies of becoming Pokémon trainers finally to have a platform where they can fulfil their dreams in reality through the telepresence medium' (Tang, 2017: 728). This points to the importance of the poetic, or story and characters, behind the game. When there is a touchpoint for nostalgia then the other interesting elements of the game, such as going outside, interacting with the real world, engaging with other players and travelling to new places, can be enjoyed (Zsila *et al.*, 2017). This nostalgia for 'beloved media characters' cannot exist when there are none connected with the game (Bonus *et al.*, 2018). It must also be remembered that a good story can overcome shortcomings in the mechanics of play and the aesthetics of the presentation (Williams, Chapter 3, this volume). Pokémon GO can serve as a guide, or template, for future location-based augmented reality games played in the physical world but connecting players with media properties, stories and characters that are already known is imperative (Vaterlaus *et al.*, 2018). Niantic, the creators of Pokémon GO and its predecessor *Ingress* have announced they are working on new game based on the *Harry Potter* universe that has enjoyed massive success on the page and screen (Niantic, 2017) and was stated as a preference over Pokémon in one piece of research (Vaterlaus *et al.*, 2018).

The third reason that Pokémon GO is not the holy grail for the gamification of tourism experiences is the fact that Pokémon GO players travel to play – collect Pokémon, interact with other Pokémon trainers and battle

Pokémon – rather than play to travel, they chase the game experience rather than a travel experience. They want to socialise, engage, cooperate and compete with others playing the game but it is about playing the game and not visiting the place. The game is the most important part of the equation and Pokémon GO players can be best described as fitting the psychological profile of gamers (Tabacchi *et al.*, 2017). The specific destination, attraction or enterprise where the game is played is a side issue. This research has shown that exposure to these entities does result in knowledge gain about the entities and this is a bonus when considering the objective of the travel. This may be an existential question of what tourism is, but there are economic advantages of people coming to a place to play while there may be social, cultural and environmental disadvantages. Pokémon GO demonstrates a way to get gamers to travel and provides a starting point for similar augmented reality applications as games. It is not the holy grail for the gamification of tourism experiences and mass tourism but it does provide a significant guidepost.

Conclusion

This chapter establishes that there are positive outcomes for tourism through playing Pokémon GO. It also shows that Pokémon GO is a form of serious leisure and introduces the idea of transitional leisure to a spectrum that ends with casual leisure. Pokémon GO is played across this spectrum and player travel is characterised by their approach to the game. Pokémon GO players are the vanguard of a gamer-based tourism niche that travels to play the game and not necessarily experience destinations and their attractions. Pokémon GO can be a template for other games that require people to travel but it is not appropriate for every circumstance in the gamification of tourism experiences.

This study found that game performance is associated with the positive aspects of Pokémon GO play and tourism, and shows that this game play is a form of serious leisure. These players are a unique group of tourists who want to play their way through destinations and attractions. The lesson to be learned from Pokémon GO is that there is a potential group of tourists who are gamers. They are willing to play in the real world where they may need to travel to and encounter new places and people. They are looking for gamified experiences that are well-designed and entertaining in terms of game mechanics, story and the opportunity for social interaction. And perhaps, like Pokémon GO, these gamified experiences need to have a connection with pop culture or a transmedia property that makes elements of the experience familiar and even nostalgic (not unlike the global network of Hard Rock Cafes). These potential tourists are in search of the serious leisure afforded by games that are both challenging, interesting, social and provide the opportunity for achievement and self-evaluation.

Practical applications for destinations

Aluri (2017: 67) suggested that tourist destinations develop customised AR games 'to the point that travelers are so captivated by the experience that they keep coming back, bringing others with them.' Successful game development is not a simple matter of following the design steps and then achieving success. A new game for a specific destination or attraction can appropriate game mechanics, aesthetics and structure from Pokémon GO but someone must build the content and make it work with the game play elements. Ultimately this new game must compete with every other new products in the marketplace. It is not guaranteed to be successful even if it does feel and play like Pokémon GO. Pokémon GO is an international game played locally and not a local game played internationally. Aluri (2017) also proposes that Pokémon GO or a similar mobile AR game can be used as a travel guide. The suggestion is to add real-world information about the location of every PokeStop and Gym so that Pokémon GO players will have access to this additional information. While this is an excellent idea for adding tourism value to the game, it is valid to question whether this strategy alone will attract more visitors to a destination, enhance engagement at the destination or improve the experience in any way.

Xu *et al.* (2016) point out that a game can only be 'attractive and desirable' when the experiential and informational motivations of tourists as game players are fully understood. This is an excellent starting point for any destination marketing organisation that wants to use mobile AR gaming to attract or engage visitors. Looking at Pokémon GO players it is important to understand that they have stayed faithful to the content over the various generations, iterations, of the games, films and TV series (Assunção *et al.*, 2017); therefore, the best thing to do with an understanding of Pokémon GO is to attract people who play Pokémon GO, starting domestically and then expanding internationally. If the enticement is good enough within the game, then the engaged player might bring a friend along for the adventure.

There are a number of things that a destination or attraction can do to entice Pokémon GO players. The first is the simple purchase of lures within the game. Lures are magnets for Pokémon that work for a limited period of time. Attracting Pokémon attracts Pokémon trainers. Churches in the USA have used lures to attract players to their location (Drescher, 2016). Cultural or heritage attractions can also use lures to attract players and then use the player presence to inform them about the location or attraction. Destinations can work with Niantic, the game's creator, to arrange the availability of special Pokémon at announced times or become a sponsor in the same way as McDonald's or Starbucks so that people know that this destination is ready for the new release of Pokémon or at least have a few interesting ones.

To effectively use Pokémon GO for its positive effects and potential to attract visitors the game itself should be embraced exactly for what it is. It is not reasonable to imagine that the success of Pokémon GO can be easily replicated. The experience gained in this endeavour to exploit the strengths of this game can then lead to the design of games and gamified experiences that are more appropriate for the destination and for the people who might want to experience those games, but it will take time. It will also be an opportunity to create relationships with the people who know how to design and make successful mobile AR games and to appropriately add game elements to an existing experience or attraction.

References

Aluri, A. (2017) Mobile augmented reality (MAR) game as a travel guide: Insights from Pokémon GO. *Journal of Hospitality & Tourism Technology* 8 (1), 55–72. DOI: 10.1108/JHTT-12-2016-0087.

Amselle, N. (2016) Pokémon GO and what it means for parks. *Parks & Recreation* 51 (8), 15–16.

Assunção, C., Brown, M. and Workman, R. (2017) Pokémon is evolving! An investigation into the development of the Pokémon community and expectations for the future of the franchise. *Press Start* 4 (1), 17–35.

Balmford, A., Clegg, L., Coulson, T. and Taylor, J. (2002) Why conservationists should heed Pokémon. *Science* 295 (5564), 2367.

Bonus, J.A., Peebles, A., Mares, M. and Sarmiento, I.G. (2018) Look on the bright side (of media effects): Pokémon Go as a catalyst for positive life experiences. *Media Psychology* 21 (2), 263–287. DOI: 10.1080/15213269.2017.1305280.

Buhalis, D., Harwood, T., Bogicevic, V., Viglia, G., Beldona, S. and Hofacker, C. (2019) Technological disruptions in services: Lessons from tourism and hospitality. *Journal of Service Management* 30 (4), 484–506.

Chung, N., Han, H. and Joun, Y. (2015) Tourists' intention to visit a destination: The role of augmented reality (AR) application for a heritage site. *Computers In Human Behavior* 50, 588–599. DOI: 10.1016/j.chb.2015.02.068.

Colley, A., Thebault-Spieker, J., Lin, A.Y., Degraen, D., Fischman, B., Häkkilä, J., Kuehl, K., Nisi, V., Nunes, N.J., Wenig, N. and Wenig, D. (2017) The geography of Pokémon GO: Beneficial and problematic effects on places and movement. *Proceedings of the SIGCHI Conference on Human Factors in Computing Systems*, 6–11 May, Denver, CO.

Cristina-Portugal, A., Campos, F., Martins, F. and Melo, R. (2017) Understanding the relation between serious surfing, surfing profile, surf travel behaviour and destination attributes preferences. *European Journal of Tourism Research* 16, 57–73.

Curtin, S. (2010) The self-presentation and self-development of serious wildlife tourists. *International Journal of Tourism Research* 12 (1), 17–33.

Domahidi, E., Festl, R. and Quandt, T. (2014) To dwell among gamers: Investigating the relationship between social online game use and gaming-related friendships. *Computers in Human Behavior* 35, 107–15.

Drescher, E. (2016) A virtual faith. *America* 215 (10), 22–25.

Frommer, D. (2017) Millions of Americans still playing Pokémon GO every day. *Recode*. Available at: https://www.recode.net/2017/3/23/15039626/pokemon-go-daily-users-chart (accessed July 2017).

Garau, C. (2014) From territory to smartphone: Smart fruition of cultural heritage for dynamic tourism development. *Planning Practice & Research* 29 (3), 238–255. DOI: 10.1080/02697459.2014.929837.

Green, B.C. and Jones, I. (2005) Serious leisure, social identity and sport tourism. *Sport in Society* 8 (2), 164–181. DOI: 10.1080/174304305001102010.
Gretel-Kauffman, S. (2016) Pokémon GO players rediscovering their neighborhoods' histories. *Christian Science Monitor*, 25 July.
Hedley, N.R., Billinghurst, M., Postner, L., May, R. and Kato, H. (2002) Explorations in the use of augmented reality for geographic visualization. *Presence: Teleoperators & Virtual Environments* 11 (2), 119–133.
Huiwen, G., Hassink, R. and Maus, G. (2017) What does Pokémon GO teach us about geography? *Geographica Helvetica* 72 (2), 227–230. DOI: 10.5194/gh-72-227-2017.
Jaramillo, G.E., Quiroz, J.E., Cartagena, C.A., Vivares, C.A. and Branch, J.W. (2010) Mobile augmented reality applications in daily environments. *Revista Eia* 14, 125–134.
Jung, T., Chung, N. and Leue, M.C. (2015) The determinants of recommendations to use augmented reality technologies: The case of a Korean theme park. *Tourism Management* 49, 75–86. DOI: 10.1016/j.tourman.2015.02.013.
Kourouthanassis, P., Boletsis, C., Bardaki, C. and Chasanidou, D.D. (2015) Tourists responses to mobile augmented reality travel guides: The role of emotions on adoption behavior. *Pervasive & Mobile Computing* 18, 71–87.
Lalot, F., Zerhouni, O. and Pinelli, M. (2017) 'I wanna be the very best!' Agreeableness and perseverance predict sustained playing to Pokémon Go: A longitudinal study. *Games for Health* 6 (5), 271–278. DOI: 10.1089/g4h.2017.0051.
Lazzaro, N. (2009) Understand emotions. In C. Bateman (ed.) *Beyond Game Design: Nine Steps Toward Creating Better Videogames* (pp. 3–48). Boston, MA: Charles River Media.
Loveday, P. and Burgess, J. (2017) Flow and Pokémon GO: The contribution of game level, playing alone, and nostalgia to the flow state. *E-Journal of Social & Behavioural Research in Business* 8 (2), 16–28.
Makkonen, T. and Hokkanen, T.J. (2013) ICT innovation and local economy: Mobile game as a tourist attraction. *Scandinavian Journal of Hospitality & Tourism* 13 (3), 257–268. DOI: 10.1080/15022250.2013.772770.
McCall, R. and Braun, A. (2008) Experiences of evaluating presence in augmented realities. *PsychNology Journal* 6 (2), 157–172.
McCall, R., Wetzel, R., Löschner, J. and Braun, A. (2011) Using presence to evaluate an augmented reality location aware game. *Personal & Ubiquitous Computing* 15 (1), 25–35.
McIntosh, E. (2016) Pokémon GO is everything wonderful about tourism. *Polygon*. Available at: https://www.polygon.com/2016/7/19/12223556/pokemon-go-tourism-nintendo-niantic (accessed July 2017).
Mesáro, P., Mandičák, T., Mesárová, A., Hernandez, M.F., Krák, B., Sido, C. and Delina, R. (2016) Use of augmented reality and gamification techniques in tourism. *E-Review of Tourism Research* 2 (1/2), 366–381.
Mortara, M., Catalano, C.E., Bellotti, F., Fiucci, G., Houry-Panchetti, M. and Petridis, P. (2014) Learning cultural heritage by serious games. *Journal of Cultural Heritage* 15 (3), 318–325. DOI: 10.1016/j.culher.2013.04.004.
Moularde, J. and Weaver, A. (2016) Serious about leisure, serious about destinations: Mountain bikers and destination attractiveness. *Journal of Sport & Tourism* 20 (3/4), 285–303. DOI: 10.1080/14775085.2016.1164069.
Nawijn, J. (2010) The holiday happiness curve: A preliminary investigation into mood during a holiday abroad. *International Journal of Tourism Research* 12 (3), 281–290. DOI: https://doi.org/10.1002/jtr.756.
Newzoo (2016) Analysis of Pokemon GO: A success two decades in the making. *Newzoo*. Available at: https://newzoo.com/insights/articles/analysis-pokemon-go/ (accessed July 2017).

Niantic (2017) The magic of Harry Potter is coming to a neighborhood near you. Available at: https://nianticlabs.com/blog/wizardsunite/ (accessed November 2017).

Parker, J. (2016) When the world is an arcade. *Atlantic* 318 (4), 40–42.

Pedrana, M. (2014) Location-based services and tourism: Possible implications for destination. *Current Issues In Tourism* 17 (9), 753–762. DOI: 10.1080/13683500.2013.868411.

Rodríguez-Serrano, A., Gil-Soldevila, S. and Martín-Núñez, M. (2017) Ludologic design and augmented reality. The game experience in Pokémon GO! (Niantic, 2016). *Revista Latina De Comunicación Social* 72, 667–678.

Scott, D. (2012) Serious leisure and recreation specialization: An uneasy marriage. *Leisure Sciences* 34 (4), 366–371. DOI:10.1080/01490400.2012.687645

Slak-Valek, N. and Williams, R.B. (2016) Pokemon GO: The effect of play on travel attitudes, social life and physical activity. Paper presented to the Travel and Tourism Research Association Asia-Pacific 2016 Conference, 15–16 November, Dubai.

Smith, C. (2017) 80+ Incredible Pokémon GO statistics and facts. *DMR*. Available at: http://expandedramblings.com/index.php/pokemon-go-statistics/ (accessed July 2017).

Stebbins, R.A. (1982) Serious leisure: A conceptual statement. *Pacific Sociological Review* 25 (2), 251–272.

Stebbins, R.A. (1992) *Amateurs, Professionals, and Serious Leisure*. Montreal: McGill-Queen's Press.

Stebbins, R.A. (1997) Casual leisure: A conceptual statement. *Leisure Studies* 16 (1), 17–25. DOI: 10.1080/026143697375485.

Suni, J. (2017) Willingness to travel as an extension of leisure activity seriousness – A study about Finnish hunters. *Journal of Outdoor Recreation and Tourism* 20, 77–83. DOI: 10.1016/j.jort.2017.10.003.

Tabacchi, M., Caci, B., Cardaci, M. and Perticone, V. (2017) Early usage of Pokémon Go and its personality correlates. *Computers in Human Behavior* 72, 163–169.

Tang, A.K. (2017) Key factors in the triumph of Pokémon GO. *Business Horizons* 60 (5), 725–728. DOI: 10.1016/j.bushor.2017.05.016.

Vaterlaus, J.M., Frantz, K. and Robecker, T. (2018) Reliving my childhood dream of being a Pokémon trainer: An exploratory study of college student uses and gratifications related to Pokémon Go. *International Journal of Human-Computer Interaction* 35 (7), 596–604. DOI: 10.1080/10447318.2018.1480911.

Wandera (2017) Pokémon GO figure: A data analysis of the most popular game of all time. *Wandera*. Available at: https://www.wandera.com/blog/pokemon-go-data-analysis-popular-game/ (accessed July 2017).

Williams, R.B. and Slak-Valek, N. (2019) Pokémon GO: The impact of an augmented-reality mobile-game on domestic and international tourism. *Information Technology and Tourism* 21 (4), 515–553.

Wu, T., Scott, D. and Yang, C. (2013) Advanced or addicted? Exploring the relationship of recreation specialization to flow experiences and online game addiction. *Leisure Sciences* 35 (3), 203–217. DOI: 10.1080/01490400.2013.780497.

Xiangyou, S.S. and Yarnal, C. (2010) Blowing open the serious leisure-casual leisure dichotomy: What's in there? *Leisure Sciences* 32 (2), 162–179. DOI: 10.1080/01490400903547179.

Xu, F., Tian, F., Buhalis, D., Weber, J. and Zhang, H. (2016) Tourists as mobile gamers: Gamification for tourism marketing. *Journal of Travel & Tourism Marketing* 33 (8), 1124–1142. DOI: 10.1080/10548408.2015.1093999

Yang, C. and Liu, D. (2017) Motives matter: Motives for playing Pokémon Go and implications for well-being. *Cyberpsychology, Behavior & Social Networking* 20 (1), 52–57. DOI: 10.1089/cyber.2016.0562.

Yovcheva, Z., Buhalis, D. and Gatzidis, C. (2012) Overview of smartphone augmented reality applications for tourism. *E-Review Of Tourism Research* 10 (2), 63–66.

Yovcheva, Z., Buhalis, D., Gatzidis, C. and van Elzakker, C.M. (2014) Empirical evaluation of smartphone augmented reality browsers in an urban tourism destination context. *International Journal of Mobile Human Computer Interaction* 6 (2), 10–31. DOI: 10.4018/ijmhci.2014040102.

Yu-Lien, C., Huei-Tse, H., Chao-Yang, P., Yao-Ting, S. and Kuo-En, C. (2015) Apply an augmented reality in a mobile guidance to increase sense of place for heritage places. *Journal of Educational Technology & Society* 18 (2), 166–178.

Zsila, Á., Orosz, G., Bothe, B., Tóth-Király, I., Király, O., Griffiths, M. and Demetrovics, Z. (2017) An empirical study on the motivations underlying augmented reality games: The case of Pokémon GO during and after Pokémon fever. *Personality and Individual Differences* 119. DOI: 10.1016/j.paid.2017.06.024.

11 Playfulness and Game Play: Using Geocaching to Engage Young People's Well-being in a National Park

Sara MacBride-Stewart, Clare Parsons and Ilona Carati

Introduction

Recent studies of young people have suggested that this group have low levels of physical activity, and that young people have become disconnected from nature through their engagement with technology (Smith & Dunckley, 2018). Technologies are dominant in young peoples' lives, and efforts have been made to understand technologies that may facilitate positive relationships between the natural and the digital world (Cord *et al.*, 2015). Geocaching is a recreational activity that combines being outdoors with gameplay (i.e. GPS coordinated 'treasure hunting'), and it offers an opportunity to experience nature through physical activity, often in an educational context (Smith & Dunckley, 2018). Although research has established the benefits of leveraging the well-being of young people through geocaching and other exergaming technologies, the focus of this literature has been on physical activity because of how these technologies seek to combine videogames with physical exertion (Li *et al.*, 2018). However, exergames are also used to promote a range of other physical benefits such as strength training, balance and flexibility but, we do not yet understand young people's experiences with these technologies, and in particular, their experience of their 'gaming' element more specifically related to play or playfulness.

In educational settings, practitioners have various reasons for designing active learning opportunities that include exergaming. These include improving educational outcomes through problem-based learning (Zemke *et al.*, 2016), efficient learning, increased attention (Lisenbee, 2015) and creativity (Lund *et al.*, 2015). Staiano and Calvert (2011: 93) review of the physical, social and cognitive benefits of exergames, claims that it 'increases caloric expenditure, heart rate, and coordination'. They add that the '[p]sychosocial and cognitive impacts of exergame play may include increased self-esteem, social interaction, motivation, attention, and visual–spatial skills'. As the geocaching is a form of exergaming that is conducted outdoors, geocaching it creates a particularly interesting opportunity to explore nature-based play surrounding this new form of educational activity.

In this chapter, the focus is on geocaching promoted as an outdoor physical activity in a protected area, to differentiate it from geocaching in urban areas and/or users who set up their own geocaches. It focuses on how the recent emergence of exergaming – videogames that also can be a form of physical exercise – creates exciting possibilities for assessing how young people might be encouraged to get outdoors and engage with national parks for recreational or tourism purposes. While outdoor recreation in national parks appears to be available to some young people according to most visitor surveys (Cord *et al.*, 2015), the potential for exergaming to enable new opportunities for experiencing nature, discovering and exploring new places, broadening and promoting outdoor physical activity means that exergaming may play an important role in combining tourism provision with education and widening access (Gray, 2007). The chapter highlights how a geocaching programme that aims to achieve sustainable active engagement with green space, was experienced by young people living near to a national park, noting in particular how geocaching might engage disaffected young people through its game/play elements.

Outdoor Recreation, Tourism and Benefits of National Parks for Young People

Outdoor recreational tourism offers important opportunities for many people to directly experience the benefits that national parks provide. This particularly concerns young people living in urban areas where the access to protected areas or large natural areas promoted for recreational or nature benefits is limited, but may have significant positive effects (Cohen *et al.*, 2016; Garst *et al.*, 2001). Although national parks are an important tourism asset, outdoor nature encounters with national parks are far less analysed in recreational tourism and hospitality literature, probably because of a strong focus in large visitor surveys on addressing infrastructure needs of traditional activity-based tourism, that is, walking, cycling and horse riding (Haukeland *et al.*, 2010).

The emergence of digital and social media that have been used to promote protected areas and national parks to visitors in new and exciting ways include the Royal Albatross cam (New Zealand Department of

Conservation; www.doc.govt.nz/royalcam) and the Natural Sounds soundscapes of the North American National Parks (www.nps.gov/subjects/sound/index.htm). These have the potential to reach a wider range of 'visitors' to these sites across a global scale. Other forms of social media offer similar possibilities – such as the promotion of tourism possibilities of national parks via Twitter and Facebook along with the sharing of photographs, videos and personal GIS mapping data (i.e. Strava). These represent novel forms of digital tourism for national parks that often rely on the provision of digital maps, visitor websites, weather data, event calendars and newsletters (Koliouska & Andreopoulou, 2013). In this context, exergaming opens up new possibilities to promote outdoor recreational opportunities and destination experiences that can engage digital users in the 'real-world' opportunities of a national park (Cord *et al.*, 2015).

However, when considering the possibility of technologically-mediated forms of recreational tourism – sometimes referred to as smart tourism (Skinner *et al.*, 2018) – to inform visitor commitments or reach conclusions about new forms of users and uses of a national park, understanding the experiences and constraints on these digital users is crucial (Shores *et al.*, 2007). In addition to the problems of rural connectivity for transport along with phone and internet signal, are the considerations of those groups with limited access to national parks as well as their perceived lack of interest in nature (Shanahan *et al.*, 2014). The use of a national park for outdoor recreational activity is not only a consumptive practice (i.e. nature as a resource to benefit mental health or social relationships, or to satisfy a need for adventure); it is also a social justice practice that can address the inequitable distribution of opportunities to access nature across social groups. Socioeconomic variation can explain some of this inequality but there has been relatively limited consideration of how nature experience varies across society (Shanahan *et al.*, 2014). Understanding young people as digital users, tells us little about the particular outdoor recreational needs of young people (Cord *et al.*, 2015). The national park site therefore creates an interesting opportunity to explore the everyday perspectives of nature-based activity as well as the possibilities of exergaming for engaging traditionally marginalised visitor groups.

Health, Outdoor Space and Young People

The disengagement of young people from physical activity and natural environments has been widely reported in the health and youth studies literatures (Ergler *et al.*, 2016). This has been linked to social changes such as digitalisation, consumption and urbanisation that are said to have generated deep changes in young people's relationships with nature (Magaudda, 2012). These changes are tied up with discourses that link increases in time spent looking at and interacting with screens relative to the time spent outdoors, to disengagement with the natural world (Smith & Dunckley, 2018).

In health studies literature, attempts to address these issues have largely been focused on ensuring that young people are able to meet physical activity guidelines as a preventative measure against other chronic conditions (Battista & West, 2018). A widely accepted discourse is that young people should be provided with opportunities to improve their levels of physical activity, with reported physical activity benefits including improvements in self-esteem and mood, enhanced concentration and cognitive performance (Jackson *et al.*, 2013). In the context of improving relationships to the outdoors, *outdoor* adventure education (OAE) in particular, is argued to improve teenagers' sense of 'social belonging, self-confidence, self-efficiency' and overall behaviour (Richmond *et al.*, 2018: 1–2).

However, evidence from the social sciences suggests that social and economic factors – rather than just personal ones – must be considered in relation to the high rates of physical inactivity and obesity reported in young people. Levels of deprivation influence access to and use of green space, with those living in the greenest quintiles of England more likely to conform to physical activity guidelines, compared with those living in the least green areas (Flowers *et al.*, 2016, Babey *et al.*, 2007). This finding is consistent for young people, as 'teenagers living in disadvantaged neighbourhoods lacked access to parks that they considered safe and therefore were less likely to participate in physical activities than teens in more affluent neighbourhoods' (Lee & Maheswaran, 2011: 217). However, where young people have good access to attractive green space, they are likely to use it, with young people reported to engage in more physical activity when in green space than elsewhere (Foster *et al.*, 2009). Maas *et al.* (2006) suggest that young people benefit more from green space than other groups, but notably this trend shows a significant decline in later adolescence.

Despite the reported benefits, it is not yet clear what kind of outdoor space is best for young people. Protected areas are said to provide greater benefits than urban parks, because they constitute a restorative environment, offer a sense of curiosity or interest and a sense of being away (Hartig *et al.*, 2014; Maller *et al.*, 2006). Protected areas also offer the highest quality opportunities for viewing natural scenes, having contact with plants and animals, and experiencing biodiversity (Romanangosa *et al.*, 2015). As such, large, biodiverse natural environments are presented as providing the most favourable environmental, aesthetic and recreational conditions for well-being (Edwards & Smith, 2012), despite being at some distance from where people reside (Kyttä *et al.*, 2015).

Consequently, there is a gap in the research literature for understanding what this means for young people, particularly owing to a concern that young people are particularly susceptible to exclusions from public space in rural and protected areas (Matthews *et al.*, 2000). In some areas, teenagers may face hostile attitudes about their use of green space because of 'an inferred association between teenagers and vandalism and crime in public space' (Lee & Maheswaran, 2011: 218; Richardson *et al.*, 2010). For young people, protected areas or national parks may be regarded as particularly inaccessible due to distance and transport requirements, money and time costs of travel, unfamiliarity and perceived exclusivity (Li *et al.*, 2018). Significantly, for young people, barriers such as major roads are a factor (Lee & Moudon, 2008). In addition, there are particular concerns about the appropriateness of protected areas for older youths who may only be considered in relation to educational or OAE goals, which risk appearing paternalistic or tokenistic (Lee & Maheswaran, 2011).

As Clarke *et al.* (2014: 200) point out, most research has 'treated natural space as homogeneous with little or no examination of the variation in health benefits according to their environmental characteristics'. There is also a need to understand how environmental and social conditions shape the health of different social groups in different ways (MacBride-Stewart *et al.*, 2016). Furthermore, based on previous research that suggests that when young people get to act as tourists, there is a need to consider the considerable linkages between their leisure and holiday behaviour rather than separate from it (Carr, 2002). When considered together, the findings suggest that the associations between physical health and outdoor space are stronger for young people who are affluent and who have good access to quality green space (Van den Berg *et al.*, 2014), and that young people's tourism behaviours are shaped by a range of sociocultural norms and values that are present in their recreation and digital lives (Carr, 2002). Thus, the experiences of marginalised young people in protected areas are important for understanding how to facilitate their active use of greenspace, in an educational and a tourism context.

The Role of Gaming Technologies in Reconnecting Young People to the Natural Environment

Exergames are part of the new generation active videogames such as Wii and Pokémon GO that link gaming technologies with exercise, through a process known as 'gamification' (Daley, 2009: 763). Shown to be effective in delivering light-to-moderate exercise, balance, strength and flexibility (Oh & Yang, 2010), the 'gamification' of physical activity through exergames refers to the integration of game-like activities into these non-game-like contexts (Huotari & Hamari, 2017). Exergames can be particularly useful for supporting young people to exercise when physical activity is combined with digital technologies that direct young people outdoors (Battista & West, 2018). The concept of gamification includes in its assessment of the effectiveness of exergames, an appreciation of the players' experience of the game, which is also a measure of its value (Huotari & Hamari, 2017). Gamification of education often leads to positive reviews of the learning experience, where digital gameplay provides more pleasure and motivation than traditional methods (Huotari & Hamari, 2017; Staiano & Calvert, 2011).

Geocaching is one form of exergaming that is growing in popularity (O'Hara, 2017; Schlatter & Hurd, 2005). Described as a 'high-tech treasure hunt that uses GPS-enabled technology to locate hidden caches' (Garney *et al.*, 2016: 1), geocaching takes gaming technologies outdoors – combining it with more traditional forms of exertion, like walking (Robinson & Hardcastle, 2015). Consistent with the claims made about exergaming more generally, there is evidence that geocaching offers a particularly enjoyable, viable alternative to walking, making it easier to motivate young people to be active (Battista & West, 2018). Its similarities to orienteering are evident in geocaching's problem-solving and discovery elements alongside positive improvements in physical activity (Chuan-Xia, 2005; Creagh *et al.*, 1997; Wood *et al.*, 2014; Xu, 2004). Like orienteering, geocaching requires participants to follow a compass arrow (or 'map') through the natural features of the landscape. However geocaching's distinctive element involves the searching and finding of treasure-caches where players can swap small trinkets or objects. When additional information and problem-solving games are linked to caches, geocaching can facilitate learning about the heritage or ecological aspects of an area (Rosier *et al.*, 2011). Less well documented are its positive psychosocial effects that come from opportunities for social interaction and relaxation in nature (Garney *et al.*, 2016).

The value of geocaching in protected areas, including national parks, has been demonstrated across several regions where it has been mainly used to promote environmental and heritage education (Schaal & Lude, 2015; Schneider & Jadczaková, 2016), as well as ecotourism opportunities (Zbuchea, 2018). In this context it has been found to facilitate understanding of natural resources (Rosier *et al.*, 2011; Santos *et al.*, 2012) and build social relations (Skinner *et al.*, 2018). The potential of geocaching to engage novice and experienced gamers actively in public, outdoor space (Lund *et al.*, 2010), makes it a valuable resource in areas seeking outreach and access opportunities for local communities (Boys *et al.*, 2017; Jacobson *et al.*, 2015). Once caches are established, geocaching can become a low-cost form of community recreation and tourism (Battista & West, 2018).

There is also a small and emerging body of literature on the potential for geocaching for opening up the tourism opportunities of natural environments to Millennials and Gen Zers (Boulaire & Hervert, 2012; Skinner *et al.*, 2018). This research suggest that geocaching may meet the needs of a new generation of tourism 'consumers' who tend to seek rich digital or gamified tourism experiences, but who also have been found to value the shared social practices that constitute gaming (Skinner *et al.*, 2018). It is also argued to have the potential to attract geocachers as tourists in ways that align with interests that are present in younger consumer's everyday lives (Boulaire & Hervert, 2012).

Much of the interest in geocaching is in the way the GPS technology can facilitate walking and exploration by 'gaming' them. However, the technology can sometime fail owing to limited signal or power, and caches can be hard to find due to the lack of specificity of GPS information below 2 metres (Robinson & Hardcastle, 2015). Furthermore, newer smartphone GPS technologies such as Pokémon GO highlight the challenge of sustaining engagement as the popularity of a game fades (Battista & West, 2018). For younger people so tied to technology, poor understanding of their digital needs and the infrastructure needed to support this can raise additional barriers to use (Ramsay *et al.*, 2017).

Despite the references in the geocaching and exergaming literature to 'game*play*', play is largely unaddressed in the discussion about physical activity, technology or young people. The focus on play is more about engaging in geocaching to improve the value of the participation, such that walking or education occur because the geocaching makes them enjoyable (Huotari & Hamari, 2017). With research suggesting that young people desire experiences in nature even if they are not engaged in them, the element of 'play' may represent a means of understanding a paradoxical problem; that in trying to reach a technologically driven and nature-curious population their motivations for nature or adventure may be different in comparison to other generations of visitors to a national park (Ramsay *et al.*, 2017). This is consistent with the large psychological literature on play that regards play as an action leading to something else – for instance throwing a ball is about learning social skills and turn taking. However, critical childhood theorists such as Burman (2016) are cautious about such claims, suggesting that they ignore the inherent merits of playing the game itself. Burman (2016) and others have called for a more child-centred analysis of play, which seeks to understand the potential of an activity to be meaningful because of how it is experienced affectively, rather than for what it does. Drawing on this, the goal of this chapter seeks to understand more about physical activity in nature through understanding how geocaching – and particularly the element of physical activity, being-in-nature and play – mediates that experience for young people. It is to conceptualising this element of play that this research now turns.

Expanding Understandings of Gameplay through 'Play'?

Within the psychologically-dominated study of play, play is regarded as a developmentally-linked and universal element of everyday childhood. Outside of these dominant discourses, play is a difficult topic to define and research. It is hard to grasp precisely what play is and does because play takes on so many different cultural and social forms, is often temporary, and its meanings and definitions change significantly across the life-course. In the context of adolescence, play is viewed as transitory and is commonly associated with risk, and because of this, the pleasurable aspects of 'play' are often lost from discourses about young people's experience of play. The possibility of geocaching offering up opportunities for game-related play appears to be a suitable route through which to grasp the concept of play for young people.

One of the dominant theoretical approaches to play has been the psychological analysis of the cognitive development of the adult, particularly regarding creativity. Underpinned by the view that play can be objectively described, this understanding has been critiqued over a lack of certainty about what aspects of the situation contributes to the creative gains of play, that is, the play itself, the interaction with others or some other element (Mellou, 1994). Mostly, such approaches say very little about the nature of play itself. Play has emerged from the literature as a complex social activity which is diverse in form and character. Elkonin (2005: 21) notes that play can be simultaneously described as: 'exercise, pleasure, relaxation, liberation', but they conclude that play is best understood as a subjective pleasure. The affective nature of play they argue, comes from the intersection of acts (i.e. gaming), objects (i.e. GPS device), spaces (i.e. nature) and relations (i.e. other gamers). It is in this intersection that play becomes meaningful to its players.

This affective understanding of play as an emotive experience with the potential to engage learning or physical activity holds particular promise for its use in understanding how the play element of geocaching can engage young people actively in nature, and in tourism-promoted nature spaces. Here, attention is given to how the play and playfulness can generate positive meanings and social connections that are associated with well-being. This approach generates an experience of well-being that is not easily measured by traditional approaches, moving beyond traditional biomedical and public health-orientated approaches which see the environment as a resource for health, towards a focus on the relationship between the feelings, place and practice that can deepen one's relationship to a place or practice. Overall, in presenting the conceptual understanding of play, based on an understanding of the practices of geocaching as a strategy for engaging young people in a national park, may contribute to our understanding of how gamified technologies can enhance this as a destination experience that meets the needs of this traditionally under-represented group.

Methodology

The geocaching project selected for study was run by the Brecon Beacons National Park Authority (BBNPA) in a funding partnership with Sports Wales. The project was designed as part of a wider programme, Calls 4 Action, addressing physical inactivity, and the generation of a positive action to address known areas of inequity and encourage new approaches to 'sport' (Gibson *et al.*, 2018). The project was aimed at local young people (11–25 year olds) in deprivation, who record some of the lowest rates of physical activity, as recorded by Sports Wales School Sports Survey (2014). The secondary aims of the project were about improving access to and usage of green space to build 'non-cognitive' skills, such as self-confidence and leadership (Richmond *et al.*, 2018: 1–2).

In this study, the participants were 62 young people enrolled in a geocaching activity aged 11–25 years; 40 females and 22 males accompanied by their parents, carers and leaders from a range of youth and community groups including those supported by Communities First. Communities First (2001–2018) was a Welsh Government community-focused, capacity-building funded programme to support the most disadvantaged people in the most deprived areas of Wales. In addition to low levels of physical activity, within this group of young people, most had never had the opportunity to visit the national park (which was 20 minutes by car), nor had they had used GPS or maps for walking, nor had access to suitable footwear and waterproofs. The Calls 4 Action funded two BBNPA activity leaders who supported each of the geocaching sessions, as well as footwear, wet weather clothing, 'safety', transport to and from the venue and GPS devices used to 'find' the geocaches and to navigate the route, as well as training for community group leaders to enable later replication. These provisions highlight the facilitative role in supporting young people to overcome immediate barriers to their access of the park.

Data was collected over six separate days with different groups. There were three phrases to the geocaching project: (1) training local youth leaders; (2) facilitated geocaching in the national park; and (3) geocaching in the local area. The youth worker training (phase 1) involved learning about how to set up and use a local trail as well as how to lead an activity (phase 3). The geocaching locations in the last two phases – the national park and local park – were used in the data collection. For many of the participants this was their first time in the national park (Jones *et al.*, 2008; Ross *et al.*, 2009).

The qualitative data were collected using mobile methods, also known as go-along, 'accompanied nature journey' (O'Brien & Varley, 2012) or walking interviews (Jones *et al.*, 2008). These methods have often been used in research concerning health (Carpiano, 2009), place (Jones *et al.*, 2008) and nature (Garcia *et al.*, 2012). In a traditional 'go-along' interview, the participant 'guides' the researcher through the virtual or real places in which they conduct their everyday life (Ross *et al.*, 2009). In this study, the interviews were literally conducted during the geocaching activity, with the researcher both observing and digitally recording the geocaching experiences as they unfolded (Evans & Jones, 2011).1 Interviews were carried out when walking between geocaches.

The interviews focused on the mobile, outdoor-nature of the activity (Cole *et al.*,1999; Garcia *et al.*, 2012). The style of the mobile interviews meant that weather and noise (wind, multiple voices) presented real challenges, but it also meant that the data were rich, dynamic and 'in-the-moment' (Garcia *et al.*, 2012). The aim of the interviews and observations was to capture sensory, emotional and immediate reflections on the natural environment. The young people were encouraged to talk about feeling, experiences and prior interactions with nature, and its importance to them (O'Brien & Varley, 2012). These pre-designed themes and questions were asked opportunistically, inserted into conversations that flowed between the young people as they walked between each geocache. Other questions were formulated during the geocaching reflecting the dynamic nature of activity and the changing view. Activity leaders (BBNPA) and young people asked questions of each other throughout (e.g. about the Brecon Beacons), and discussion was encouraged by other adults present (carers, parents and community workers). These recorded conversations contributed to the rich data.

The go-along interview is well-suited to capturing novel or new experiences, and observations of embodied, 'in-the-minute' encounters (Evans & Jones, 2011; Carpiano, 2009). This method is useful for recording experiences that are 'fleeting or taken-for granted' (Currie *et al.*, 2016: 643), unanticipated or unplanned (Clark & Emmel, 2010; Moles, 2009). These approaches are helpful because participants can find it easier to verbalise attitudes and feelings as they happen (Evans & Jones, 2011). In addition, the multiple voices of the group combined with the soundscape of the outdoors, gave the interviews a noisy yet dynamic quality. During the walk, as participants moved in and out of range, opting-out or including themselves in the conversation, they disrupted the usual hierarchy created between researchers and participants (Garcia *et al.*, 2012).

The study used an opt-out consent process. Ahead of the visit, the BBNPA activity leaders sent the youth workers plainly-worded research and consent information so they could advise people registering for the activity about the research, and the presence of researchers. On arriving at the site, the researcher (sometimes accompanied by a student) was introduced to the group and the process for consent was re-explained. The researchers carried a handheld recorder and initiated each discussion with an introduction and a confirmation of participation. At the beginning and end of each session, a group discussion was facilitated by the activity workers and recorded.

There was number of follow up 'static' interviews with activity workers (2), young person (1) and a youth worker (1) in the weeks immediately following the geocaching, and an additional four interviews with youth workers in a 12 month follow-up. The interviews were at the National Park Visitor Centre, the youth centre or by telephone when there were travel constraints. There were fewer follow-up interviews than originally planned, due to time constraints of youth workers and the chaotic nature of the young people's lives.

Transcription of the interviews occurred with the participants' consent and the transcribed interviews were subsequently analysed. Following Braun *et al.*'s (2014) qualitative analysis approach, the data were initially reviewed by two researchers and grouped into common-sense themes. These themes can be described as the 'how', 'what' and 'why' of geocaching that explain its health impact. The interaction of physical activity and nature (how), geocaching and nature (what) and well-being and being a young person (why) describes the initial grouping of the data into themes. Following Hollway and Jefferson's (2000) psychosocial approach, these 'common-sense' themes were re-interpreted through the theoretical lens of 'play' outlined in the literature review. This analysis sought to move the analysis away from everyday claims about what geocaching can achieve, to a deeper, richer account about how geocaching engages young people in nature and physical activity. The content of the themes was verified by returning to the literature on exergaming and green care.

A mixed-method secondary analysis of the (109) 'end-of-day' participant surveys, was also completed. The validity of these surveys has to be considered, due to age and possible low literacy. The survey data show the immediate effects of the activity, and how participants felt about their experience. The survey data were grouped together with the interview data, under common themes. The analysis is presented in the next section.

Results

Physical activity and nature

In this research, play was never directly asked about. It emerged in the observations of the geocaching and in the interviews with the activity leaders. Instead, the activity leaders asked specific questions about levels of physical activity focusing on walking in particular. Questions were asked about enjoyment and distances normally walked. The aim of these questions was to explore young people's familiarity with walking as a form of physical exercise, and what they felt about it while they were walking.

The responses could be divided into two groups: those who were reluctant about walking and whose activity levels were low, and those for whom the distance walked was familiar and who did not raise objections to walking while geocaching. The 'end-of-day' survey data found that walking was disliked by 30% of the group. This corresponded to reported everyday activity levels – 40% of the participants spent less than an hour on an outdoor activity a week, while a further 40% spent more than two hours a week. Over half of the respondents positively conveyed that the distance walked was more than they normally walked. With the geocaching generating a sense of achievement among the participants, it should be noted that the average estimate was close to double the actual distance walked, with some (males and those normally active) being slightly better at predicting distances.

In addition to questions about the physical distances walked, the young people were asked about the places, barriers and opportunities for walking or spending time in a national park. The barriers highlighted by the young people included issues with accessibility (including not having a reliable car), as well as a lack of knowledge about the existence of the national park. When asked if they would visit again, the number of young people willing to do so (39%) was higher than for those ruling out subsequent visits (11% 'no-unlikely'); with 50% 'unsure'. This uncertainty may be due to the relative dependence on adults for activities and access, leading to a lack of opportunity or understanding. Answers such as 'just haven't done any because I haven't been asked' could imply a desire to be involved in more outdoor activity and spaces, but it reinforces the suggestion that rather than being uninterested, young people are likely to lack the opportunity to be active in these spaces.

Owing to the reluctance and uncertainty about walking expressed by some young people, the youth leader's actively avoiding talk about 'walking', emphasising instead its positive aspects related to achievement, exploration and geocaching (i.e. game-play, GPS). Positively framed discourses about physical activity and being outside, filtered into the talk of young people, particularly those who were already active. As the extracts below suggest, the nexus of the environment and activity was potentially seen as 'fun', although – as shown in the second account – being outdoors was more simply an alternative to being indoors:

I: So you did it [geocaching] the first time in the rain… why did you come back?
YP: Because it's fun, because it's active. [Field04Aug]

I: So you like being outdoors? What do you like about being outdoors?
YP: Better than being stuck inside.
I: Do you mind it when it's wet though?
YP: Yeah, I don't mind when it's wet. I don't really like getting cold. [Field04Aug]

Furthermore, the experience of walking was about feeling physically capable. This sense of physical capacity acknowledges the affective as well as the physical demands of walking. That is, the sense of achieving or reaching something beyond ones' own expectations. The youth workers linked this discourse of achievement to the improved confidence of the young people following the geocaching:

YW: It was quite a big deal. A big thing has been improved confidence as a result...To have walked all morning is a really big deal. [IntYouthWorker1]

YW: In the bus going home (laughs), all agreed their level of fitness was pretty low... the immediate impact was you know, going from an exercise and fitness point of view [IntYouthWorker2]

YW: It really is building on that self-esteem and that confidence that you can do it and nobody's going to laugh at you if you fail. You know they are able to risk failure and still be able to um cope with it [IntYouthWorker2]

In addition to the emotional and physical aspects, the geocaching was viewed as having a social nature. This was evident in interactions that young people had that were outside of their comfort zone, as well as conversations with people they would not normally talk to. The activity leaders commented on the improved relationships within the groups that emerged from the sense of social connectedness and a shared experience. This account contradicts a more dominant view in the literature on gaming, that gaming technologies produce, rather than resolve, social isolation (Staiano & Calvert, 2011).

YW: Our pupils are quite used to being indoors and the thought of going outside is quite a scary, you know, scary um experience for them, but having something to occupy their mind while they're out there doing it, makes a huge difference [IntYouthWorker4]

Furthermore, the description of the landscape reported here is helpful for understanding what walking was like for these young people, and the challenge that walking presented to them. Particularly for those young people who did not play sports, environmental factors (i.e. getting wet), laziness, a lack of interest and possible fears ('because you do dangerous things') were given as reasons for disliking the outdoors. However, a sense of overcoming the potential hurdles of uneven, unfamiliar terrain and open space contributed to a positive experience of walking.

YW: She said she didn't realize she could do that walking, she wouldn't have thought she was up to it and that she actually enjoyed it. It's a major hurdle I hadn't thought of. It can be something that simple. Just distance walk for some of them, and the fact the environment it wasn't pavement you know, just open, uneven ground, no obvious paths in places you know, that was very new to a lot of them [IntYouthWorker1]

For these young people, a discourse of being or becoming physically capable incorporates the physical, emotional and social challenges of the landscape in which the walking is done. The youth workers often commented on the long-term benefits of these effects that included positive behavioural changes that extended for several days after the geocaching. However, it was impossible to ascertain whether the reported activity was sustained by the youth groups or by individuals. Ideally the study hoped to follow-up participants after 12 months, but the precariousness of the young people's lives and the insecurity of youthwork as a job meant that only a small number of these interviews were completed.

While young people experienced the walking element of geocaching positively, it was service providers who dominated the discussion about its value for physical activity and social connectedness (Fox & Alldred, 2015). Notably the importance of the social value of the national park space to younger people emerged from their responses. Young people seemed less interested in the physical benefits but were more interested in what the walking – and the environment in which they walked – *felt* like. It is to this sense of the environment or nature that this analysis now turns.

Geocaching and nature

While the focus of geocaching research has been on the qualities or skills that it enhances, for example balance and flexibility (Oh & Yang, 2010), there has been little attention to what kind of environment or experience the geocaching is trying to 'game', that is, the role that exergames have in surreptitiously trying to engage young people in physical activity and in nature spaces. The ways in nature was perceived by the young people – and the ways in which nature is mediated by the geocaching – are addressed in this theme. This has relevance for understanding how gamification might enhance the destination experiences of young people (Skinner *et al.*, 2018). The play element of geocaching and the gamification of physical activity and the outdoors, is addressed in the later sections (Huotari & Hamari, 2017; Staiano & Calvert, 2011). This theme starts with a question that the activity workers would ask at the start of the day, to find out whether the young people were familiar with the idea of a National Park, and whether they had visited the park previously.

> **AW:** They've all heard of Pen Y Fan [the highest mountain in the park] so it's really useful to stand there and go that, that's it over there. They don't realise it's that close. To them it's just sort of a mythical mountain. Somewhere that's supposed to be really high, but to see it – then sort of try, I mean, just try...you know, sort of attitude. [IntActivityworker1]

When first asked about their knowledge of the national park, the young people were most likely to talk about it in terms of its iconic status, as somewhere that they wanted to visit one day or to climb its highest mountain. Mythical places are common in gaming narratives, but in this context, the landscape, its views and terrain had important cultural, aesthetic and symbolic qualities that materialised once the young people arrived in the national park. At the same time, these symbolic aspects existed in the absence of any other formal relationship to the national park. The young people knew this was an important place, but few had been there or knew it was so close to them. The physical proximity of the park to their homes was often a surprise to the young people. If young person had already walked this mountain, it had been with the support of a youth or sports group, and in a rare case, a parent. Of course, a key aim of the geocaching project was to facilitate the young people's relationship to nature by 'getting young people outdoors'. Without knowing that the geocaching was 'gaming' being outdoors in order to generate more positive experiences (Staiano & Calvert, 2011), the parents who accompanied the young people, reiterated this point:

AW: Gets you outdoors. Not stuck in the house in front of the TV. Fresh air you know...Kids enjoy it! 'Look at that look at that!' They don't do that now do they? They don't play in the woods and stuff like that. They enjoy it....And they're still learning. They've never used one of these before. [Field09Aug]

Literature on outdoor education or therapeutic environments has a tendency to assume that getting outdoors is beneficial and desirable for young people, mainly because it is linked to discourses about using nature as a playground. For many of these young people – many of whom had never visited a national park before – nature instead appeared confusing. This confusion was evident in the language and descriptions they used. They often used the wrong terms for animals and plants, not able to distinguish between a bramble and a nettle for example, or unable to list or label trees when asked. Their experience of the natural world was often negative, as evident is expressions like 'I don't like it' when asked about being outdoors. At the same time, nature was associated with smells, with difficulty walking and scary or unfamiliar animals. It was also sometimes perceived to be dangerous and physically cold or wet:

YP: What's that?
AW: That's a slug.
YP: Urrhhhhhh. They look disgusting! [Field25July]
YW: Lot of questions that seem a little out of left field, are there bears, how big are the birds, will there be any sort of insects and that sort of thing. But that just comes from not having been in that space before. Most of them have never had a meal outdoors so having a picnic is quite a big thing as well. [IntYouthWorker]

With nature appearing to be inaccessible, and the experience of being-in-nature having the potential to distance young people from it, the activity leaders recognised that there was a need for nature to be mediated. Drawing on outdoor education skills, the activity leaders used a range of approaches that helped make this nature more accessible; for example, by encouraging talk about what animals they liked (and feared) and talk about domestic environments and pets, for instance, helped participants to connect with the environment around them. Attention was paid to how current experiences of nature were shaped by their prior experiences.

For the leaders, what mattered was providing support, being realistic about what could be achieved and being sensitive to the challenges faced by young people. Activity workers set boundaries, built trust and allowed the young people to save face, using humour and through taking on responsibility or leadership. This is consistent with work with vulnerable people that establishes a core set of values (i.e. respect, noticing) and practices (learning, listening) and principles. There was also considerable allowance in the programme design for the possibility that young people resist choices about how to exercise. Investing in material things such as individual GPS, boots and raincoats, and having trained experienced staff all increased the potential for play and enjoyment in the natural environment by reducing the anxiety or worry that shoes might get muddy or that you might get wet:

I: What did you like about [the geocaching]?
YP: All the walking we did.
I: You liked all the walking. You didn't mind the sheep poo?
YP: I wore my new trainers.
I: Oh you wore your new trainers?
YP: Yeah but I asked for wellies. I got them. I didn't mind getting them dirty. [Field30Aug]

While animals and objects were used to mediate discussions about nature, the geocaching, also had a similar role. The GPS devices enabled the young people to move through the environment in ways that would otherwise require the use of detailed maps. These devices directed users where to go – relying on direction – rather than terrain features that would have had to be interpreted. The GPS technology did not require expert knowledge and they mediated access to the space through by the introduction to the environment through 'treasure hunting' (Smith & Dunkley, 2018).

I: If you were hiding [the cache] where would you hide it?
YP: There.
I: What do you think would happen in winter? All this bracken wouldn't be here would it? It'd be really easy to see.
YP: It'll be up a tree!
I: Up a tree? I don't fancy climbing a tree. Do you climb trees?
YP: On the ground, I think. [Field09Aug]

The leaders also used the geocaching to 'teach' on diverse themes ranging from history, to archaeology, maths, geography and Welsh language as well as biology: 'So what do you think this thing was? Maybe like a defence? A fort? Or do you think it was made by nature? [Int04Aug]'. Here the natural and educational environment is used to connect the young people with a sense of local history, identity and place. These connections were made through descriptions given by the leaders, and information contained within the caches about physical objects that are encountered during geocaching; for example, a standing stone, a Roman road or a viewing spot. Unlike other gaming technologies, physical engagement was possible through touch or the taking of a photograph. The caches (which include the educational content) and the activity leaders both have an important role facilitating these connections and in extending the potential of the geocaching device. The cache locations become more than a static compass point; their game-like qualities served as points through which nature and the natural environment became discoverable and fun.

This section has described the lack of material access to, and inexperience of, the natural environment and the national park for these young people. Geocaching had a role in mediating both access and experience, with a positive outcome that appears to transform young people's negative experiences. In many ways, it literally used the winning and finding of caches as form of gaming practice that reached its intended aim of encouraging young people to be physically active in nature spaces. The gamification of nature was also used to enhance their destination experience. The success of the geocaching in achieving this was evident in questions to activity leaders by some young people wanting to know more about other trails they could use, how to get to the national park, get footwear or a personal GPS. Through geocaching then, nature and the natural environment became something that was possible to discover and be enjoyed, rather than being directly learnt about (Whatmore, 2006). Drawing on an educational and therapeutic approach, the activity leaders played an important role in facilitating a positive experience with nature, providing both support and equipment. However, less is known about the processes through which physical activity or nature is positively experienced by young people. That is, how these young people could reap social benefits from the gamification of nature, rather than the gamification of nature being the ultimate goal (Skinner *et al.*, 2018) It is to an analysis of play that this chapter now turns.

Well-being and being a young person in Wales

From the perspective of the parents and carers, play was an important element that they felt had been lost from their teenager's lives. This is consistent with the idea in the youth studies literature that young people lack physical play and opportunities to be in nature because of their engagement with digital technologies (Smith & Dunckley, 2018). While this section talks about play, the analysis avoids the technicalities of how geocaching generates gameful play; rather, play is analysed from the perspective of how this nature-based activity is experienced and what the 'play' element of geocaching achieves for this group of young people

In the interviews with parents and carers, the importance of play was linked to its physical and metaphorical representation as being about creativity and freedom, particularly in relation to being outside. This is discourse of play is supported by the psychological literature, where play is regarded developmentally as being important for helping to 'push the boundaries' and facilitate learning. The idea of a 'playful' engagement with nature and physical activity is not traditionally presented as a key geocaching benefit for young people. In contrast to orienteering, for example, which emphasises competition and problem solving, using game-thinking in geocaching was much more about its treasure seeking and exchanging objects. Here, geocaching is attuned to the role of 'playability' and 'purposefulness' for addressing young peoples' reluctance to be physically active in nature. Reflecting on this element, geocaching could be perceived to have been somewhat successful in connecting people to nature:

I: What did you like?
YP: No-one's here
YP: I dunno. It's open, I'm walking round in wellies, leggings, and I don't feel weird about it
YP: I haven't got full make up on, I've just got mascara on. That's it. The only people who're gonna see me are the farmers, the sheeps. Feel comfortable. [Field31Aug]
YW: So it's nice for em to push the barriers you know...they know they're by here and can have a run about and they're learning quite a bit without knowing it. [IntYouthWorker]

In drawing attention to play in the context of geocaching, it is not unexpected that the novelty of the nature/activity experience will give rise to playful behaviours and approaches in young people. However, playfulness emerged not only from the design elements of geocaching, but also young peoples' capacity to use the geocaching for purposes that reflected their own needs and social expectations. Although this was alluded to in the previous section, what play does and how it is experienced reflected simple, almost mundane but situational opportunities for engaging in the environment in novel ways. Certain features of the environment – the puddles, the ferns, the sheep poo – helped the groups connect with nature through their physical bodies, and often in playful ways. These nature objects were registered through both pleasure and fascination, and abject disgust; but these were often expressed through humour and a reminder not to take this being in nature too seriously.

I: Ok, what did you like best about this afternoon?
YP: Seeing sheep poo nah I'm joking!
YP: Walking in the puddles!
YP: Same. Puddles!!! [Field3108]

For many of the participants play was something that might be associated with younger children. Play, and particularly physical outdoor play, is symbolic of childhood; its symbolism was therefore affectively significant for this group otherwise regarded as being in transition. Physically placing young people in a national park, away from the structures and routines of everyday life, gave this playfulness an otherwise illicit and significant character. Furthermore, having the opportunity to be experience nature while being removed from the expectations of other peers, and the demands of adolescence, provided an opportunity for the young people to engage in a sensorial and affective way, that is, through the feel of the mud, the risk of the water penetrating the boots and the possibility of falling.

The facilitated role that the activity leaders took on was one of word play – using familiar language and language play to generate conversations about nature. Shared (playful) banter was a common way of playing for the young people. Banter in particular is a form of shared playful language that sometimes included the adults but mostly it distinguished them in order to strengthen peer-related social bonds. Combined with the recognition that young people are often rapid up-takers of new forms of language ('emojis', 'txt spk', etc.) and new words ('snake', 'buzzing'), this word play was an effective tool for engaging them with the environment. In the following extracts the young people were often willing to play along, injecting additional humour into the discussions:

AW: Anybody know any trees that they can see?
YP: Oak!
AW: That's not an oak. But an oak is a tree so that's very good. What other trees do you know?
YP: A mossy [tree]
YP: Christmas tree.
YP: Elm
AW: An Elm. Wow.
YP: A lavatory! (laughter). [Field09Aug]

AW: What is it?
YP: It's a feather!
AW: I knew that but from which bird?
YP: A black bird
YP: A hawk
YP: Magpie that's the one
AW: You're gonna be amazed

YP: Big magpie.
YP: A heron!
YP: A pheasant!
YP: An eagle!
AW: It is a bird of prey…
YP: Eagle's close.
AW: Who knows a big brown bird of prey?
YP: Me.
AW: It's not an owl feather.
YP: That's a beautiful feather.
AW: Begins with B.
[Kids all shouting bird]
YP: Something that flies!
AW: Bzzzz.
AW: This is a buzzard feather
AW: Do you know what birds of prey eat?
YP: Um, worms.
AW: No, even bigger than that.
YP: Uh, rabbits!
YP: Blackbirds!
YP: Maybe like a tiny sheep.
AW: Maybe like a baby sheep. If the sheep was injured. How about mice? Voles, stoats…
[Field04Aug]

The multidimensional character of play is reflected in symbolic, structural, sensory, linguistic and affective elements described above (see also Elkonin, 2005). This illustrates a key benefit of the gamification of nature offered by the geocaching – that it enabled various modes of engagement that were attached to the geocaching design, but also reflected the users' experiences and active enactments with nature (Huotari & Hamari, 2017). In the context of geocaching, the playful engagement in nature becomes the forefront of the experience of the environment rather than the more functional and instrumental approach to walking. However, it is also important to locate this finding in wider tourism research with younger people, which reminds us that the needs expressed by young people in a recreational or outdoor setting are more likely to reflect how they will act 'on holiday', rather than being separate from it. This supports the view that opportunities to play are likely to be as important to young people as the destination (Skinner *et al.*, 2018).

Discussion: Gaming of Nature Through Play

Gaming technologies are thought to disconnect young people from nature spaces. Yet this research has shown the usefulness of exergaming technologies in the context of a supported geocaching programme for connecting young people to nature in both direct and indirect ways (Cord *et al.*, 2015). The analysis has shown how a lack of opportunity, reluctance and uncertainty about being in nature meant that these young people were unlikely to have had regular experiences of being physically active. The geocaching resulted in a sense of achievement, often through the experience of walking and being on unfamiliar terrain. This achievement was not just about doing physical exercise but rather it was the experience of what being in nature or doing physical activity *feels* like. Building on an affective understanding of play (Elkonin, 2005), geocaching was shown to serve as an important tool for opening up pleasurable opportunities for engagement in nature and physical exercise.

The idea that play can be a powerful emotive experience that can engage young people in the natural world is supported by Elkonin's (2005) work on the subjective pleasures associated with play. In this approach, play has been described as a complex social activity that relies on the intersection of acts, objects, spaces and personal relations. The geocaching does this in a way that 'games' nature, literally by using each of those elements (the geocaching, the caches, the national park, etc.) in purposeful ways and as part of the strategy for engaging young people. In this study, playful encounters with nature occurred in part because of the provision of formal and informal sources of support that reduced the 'stress' of being outdoors, and connected the young people to a sense of local history, identity and place.

It was through play that the pleasurable benefits of being-in-nature were mostly realised. The meaningfulness of 'play' rather than 'physical activity depended to some extent on the prior experiences, perceptions and expectations of the young person. Puddle stomping, banter and fern diving – generated laughter, curiosity and interest – and it was these rather than physical activity that produced experiences normally associated with the benefits of being in a green space (see Hartig *et al.*, 2014).

In the literature, such child-centred experiences of nature are often missing (Burman, 2016). A particularly interesting finding of this study is the way in which young people also use play to actively resist attempts to take physical exercise or the experience of being in nature too seriously. Furthermore, with nature being experienced as overwhelming and dirty, the possibility that it could also be a source of fun mitigated some of its negative impacts.

Playfulness is not fully considered in the gaming literature as a central aspect of gamification. Deterding *et al.* (2011) suggest that playfulness is a complementary element of exergaming, which makes the game (i.e. the geocaching) more enjoyable – and therefore potentially more motivating to use (see also Huotari & Hamari, 2017; Staiano & Calvert, 2011). While curiosity, exploration and pleasure may be experienced in other physical activities, in exergaming the play is 'gamified', that is, it is built into the outcomes of the game. The problem with this approach is that play is limited to what it can achieve, rather than recognised as a desirable user experience or way of interacting with nature and with the outdoors. From the perspective of the users here, the exergaming element engendered a playfulness that sets the geocaching aside from other physical activities or nature encounters, or from other educational, environmental and heritage uses of geocaches with a national park context (Schaal & Lude, 2015; Schneider & Jadczaková, 2016). So, to restate, while geocaching may fill the necessary expectations of a physical activity programme and opportunities to engage young people in the destination experience, for these young people it was the playfulness and meaningfulness of that playfulness that constructs novel expectations of being in nature and physical activity.

Conclusion

The purpose of this chapter was to present an understanding of the destination experience of national parks – as a place of nature – for young people, based on a conceptual approach to the 'game' element of geocaching. Geocaching has been described as a digitally-based form of recreation useful for positively promoting being outdoors or in nature, for young people who often feel excluded from these opportunities. A key finding of this research is that geocaching and indeed gamification is a complex social activity that involves the cooperative elements of well-resourced staff (in this case specialist youth activity workers), other resources (transport, equipment) and a set of youth-centred values and practices. Rather than seeing young people as a new source of consumers for the tourism market, this research suggests that involving young people in the opportunities available from nature-based tourism requires a principled approach that reduces inequalities to access, while harnessing the wave of exergaming technology that could widen the audience for outdoor recreation.

Given that national parks remain relatively inaccessible to some young people, the lack of access to, and knowledge about, nature spaces was a very real experience for these young people. Parents and carers linked this loss to the impact of digital technologies on young people. This view is supported by academics who argue that digital forms of activity, including videogames, are a contributing factor to a lack of access to green space (proposed earlier), and a lack of physical activity in a natural setting (Björk *et al.*, 2008; Leder, 1990; Magaudda, 2013). Work in this frame notes the significance of digital technologies for negatively shaping health behaviour and experiences of nature. However, in response to such work, from this research it has been possible to see how exergaming provided an opportunity for marginalised, relatively in/active young people to experience the national park. Through playful relationships generated with others and mediated by technology, this research shows the potential for engaging young people positively in nature reflected in physical and sensorial expressions of pleasure and fun. Facilitated and supported opportunities to play, as much as learning or being active, may encourage future use, particularly for teenagers who are often excluded, or feel excluded, from such places.

Placing gamification at the heart of this analysis of geocaching offers an opportunity to think about the existing relationships that young people might have to play, particularly given the often contrived or purposeful nature of play in educational and learning settings, and its under-representation in the nature tourism literatures (Sheller & Urry, 2004). Thinking about gameplay in particular, lends its way to assessing what is needed for a young person to feel able to play, that is, accessible gaming technology, wellies, as well as the support and respect required from facilitators, and those who deliver tourism or visitor services in national parks. Bringing young people to nature and the outdoors was experienced by them as the removal of some of the traditional constraints on them, and it is accounted for here as enabling a sensorial and affective form of play. In part, the mobile methods used in this study also helped facilitate these insights.

Note

(1) Journalism-standard, hand-held recording devices (Zoon H2N) for use in outdoor conditions were used.

References

Babey, S.H., Hastert, T.A. and Brown, E.R. (2007) Teens living in disadvantaged neighborhoods lack access to parks and get less physical activity. Los Angeles: UCLACenter for Health Policy Research. https://escholarship.org/uc/item/4ww2p7v4 (accessed 5 August 2020).

Battista, R.A. and West, S.T. (2018) The use of geocaching as a form of physical activity in youth. *American Journal of Health Education* 49 (3), 125–132.

Björk, J., Albin, M., Grahn, P., Jacobsson, H., Ardö, J., Wadbro, J. and Skärbäck, E. (2008) Recreational values of the natural environment in relation to neighbourhood satisfaction, physical activity, obesity and wellbeing. *Journal of Epidemiology & Community Health* 62 (4), e2. DOI: 10.1136/jech.2007.062414

Boulaire, C. and Hervet, G. (2012) New itinerancy: The potential of geocaching for tourism. *International Journal of Management Cases* 14 (4), 210–218.

Boys, K.A., DuBreuil White, K. and Groover, G. (2017) Fostering rural and agricultural tourism: Exploring the potential of geocaching. *Journal of Sustainable Tourism* 25 (10), 1474–1493.

Braat, L. and de Groot, R. (2012) The ecosystem services agenda: Bridging the worlds of natural science and economics, conservation and development, and public and private policy. Available at: http://ac.els-cdn.com/S2212041612000162/1-s2.0-S2212041612000162-main.pdf?_tid=55343628-80db-11e7-95c6-00000aacb35d&acdnat=1502706697_ed611a445cfc2c11f0abc542ed9b8625 (accessed 9 August 2017).

Braun, V., Clarke, V. and Terry, G. (2014) Thematic analysis. *Qualitative Research in Clinical Health Psychology* 24, 95–114.

Burman, E. (2016) *Deconstructing Developmental Psychology*. London: Routledge.

Carpiano, R.M. (2009) Come take a walk with me: The 'go-along' interview as a novel method for studying the implications of place for health and well-being. *Health & Place* 15 (1), 263–272.

Carr, N. (2002) Going with the flow: An assessment of the relationship between young people's leisure and holiday behaviour. *Tourism Geographies* 4 (2), 115–134.

Chun-Xia, C. (2005) Study on the effects of orienteering on the middle school students' mental health. *Fujian Sports Science and Technology* 2, 7–19.

Clark. A. and Emmel, N (2010) Using walking interviews. Realities Toolkit #13 ESRC National Centre for Research Methods. Available at: http://eprints.ncrm.ac.uk/1323/1/13-toolkit-walking-interviews.pdf (accessed 19 July 2018).

Clark, N.E., Lovell, R., Wheeler, B.W., Higgins, S.L., Depledge, M.H. and Norris, K. (2014) Biodiversity, cultural pathways, and human health: A framework. *Trends in Ecology & Evolution* 29 (4), 198–204.

Cohen, D., Han, B., Nagel, C., Harnik, P., McKenzie, T., Evenson, K., Marsh, T., Williamson, S., Vaughan, C. and Katta, S. (2016) The first national study of neighborhood parks. *American Journal of Preventive Medicine* 51 (4), 419–426.

Cole, D., Eyles, J., Gibson, B. and Ross, N. (1999) Links between humans and ecosystems: The implications of framing for health promotion strategies. *Health Promotion International* 14 (1), 65–72.

Cord, A.F., Roeßiger, F. and Schwarz, N. (2015) Geocaching data as an indicator for recreational ecosystem services in urban areas: Exploring spatial gradients, preferences and motivations. *Landscape and Urban Planning* 144, 151–162.

Creagh, U. and Reilly, T. (1997) Physiological and biomechanical aspects of orienteering. *Sports Medicine* 24 (6), 409–418.

Currie, M.J., Lackova, P. and Dinnie, E. (2016) Greenspace matters: Exploring links between greenspace, gender and well-being with conservation volunteers. *Landscape Research* 41 (6), 641–651.

Daley, A. (2009) Can exergaming contribute to improving physical activity levels and health outcomes in children? *Pediatrics* 124 (2), 763–771.

Deterding, S., Dixon, D., Khaled, R. and Nacke, L. (2011) From game design elements to gamefulness: defining gamification. Proceedings of the 15th International Academic Mindtrek Conference: Envisioning Future Media Environments (pp. 9–15). Mindtrek11, September 28–30, 2011, Tampere, Finland.

Edwards, D., Jay, M., Jensen, F.S., Lucas, B., Marzano, M., Montagné, C. ... and Weiss, G. (2012) Public preferences for structural attributes of forests: Towards a pan-European perspective. *Forest Policy and Economics* 19, 12–19.

Elkonin, D.B. (2005) The psychology of play. *Journal of Russian & East European Psychology* 43 (1), 11–21.

Ergler, C.R., Kearns, R., Witten, K. and Porter, G. (2016) Digital methodologies and practices in children's geographies. *Children's Geographies* 14 (2), 129–140.

Evans, J. and Jones, P. (2011) The walking interview: Methodology, mobility and place. *Applied Geography* 31 (2), 849–858.

Flowers, E., Freeman, P. and Gladwell, V. (2016) A cross-sectional study examining predictors of visit frequency to local green space and the impact this has on physical activity levels. *BMC Public Health* 16 (1). DOI: 10.1186/s12889-016-3050-9.

Foster, C., Hillsdon, M., Jones, A., Grundy, C., Wilkinson, P., White, M. and Thorogood, M. (2009) Objective measures of the environment and physical activity – Results of the environment and physical activity study in English adults. *Journal of Physical Activity and Health* 6 (s1), S70–S80.

Fox, N.J. and Alldred, P. (2015) Inside the research-assemblage: New materialism and the micropolitics of social inquiry. *Sociological Research Online* 20 (2), 122–140.

Garcia, C.M., Eisenberg, M.E., Frerich, E.A., Lechner, K.E. and Lust, K. (2012) Conducting go-along interviews to understand context and promote health. *Qualitative Health Research* 22 (10), 1395–1403.

Garney, W.R., Young, A., McLeroy, K.R., Wendel, M.L. and Schudiske, E. (2016) A qualitative examination of exergame motivations in geocaching. *Games for Health Journal* 5 (1), 34–39.

Garst, B., Scheider, I. and Baker, D. (2001) Outdoor adventure program participation impacts on adolescent self-perception. *Journal of Experiential Education* 24 (1), 41--49.

Gibson, S., Loukaitou-Sideris, A. and Mukhija, V. (2018) Ensuring park equity: A California case study. *Journal of Urban Design* 1–21.

Gray, H.R. (2007) Geo-caching: Place-based discovery of Virginia state parks and museums. *Journal of Museum Education* 32 (3), 285–291.

Hartig, T., Mitchell, R., De Vries, S. and Frumkin, H. (2014) Nature and health. *Annual Review of Public Health* 35, 207–228.

Haukeland, J.V., Grue, B. and Veisten, K. (2010) Turning national parks into tourist attractions: Nature orientation and quest for facilities. *Scandinavian Journal of Hospitality and Tourism* 10 (3), 248–271.

Hollway, W. and Jefferson, T. (2000) *Doing Qualitative Research Differently: Free Association, Narrative and the Interview Method*. London: SAGE.

Huotari, K. and Hamari, J. (2017) A definition for gamification: Anchoring gamification in the service marketing literature. *Electronic Markets* 27 (1), 21–31.

Jacobson, S.K., McDuff, M.D. and Monroe, M.C. (2015) *Conservation Education and Outreach Techniques*. Oxford: Oxford University Press.

Jackson, L., Daniel, J., McCorkle, B., Sears, A. and Bush, K. (2013) Linking ecosystem services and human health: The eco-health relationship browser. *International Journal of Public Health* 58 (5), 747–755.

Jones, P., Bunce, G., Evans, J., Gibbs, H. and Hein, J.R. (2008) Exploring space and place with walking interviews. *Journal of Research Practice* 4 (2), 2–9.

Koliouska, C. and Andreopoulou, Z. (2013) Assessment of ICT adoption stage for promoting the Greek National Parks. *Procedia Technology* 8, 97–103.

Kyttä, M., Hirvonen, J., Rudner, J., Pirjola, I. and Laatikainen, T. (2015) The last free-range children? Children's independent mobility in Finland in the 1990s and 2010s. *Journal of Transport Geography* 47, 1–12.

Leder, D. (1990) *The Absent Body*. Chicago: University of Chicago Press.

Lee, A.C. and Maheswaran, R. (2011) The health benefits of urban green spaces: A review of the evidence. *Journal of Public Health* 33 (2), 212–222.

Lee, C. and Moudon, A.V. (2008) Neighbourhood design and physical activity. *Building Research & Information* 36 (5), 395–411.

Li, J., Erdt, M., Chen, L., Cao, Y., Lee, S.Q. and Theng, Y.L. (2018) The social effects of exergames on older adults: Systematic review and metric analysis. *Journal of Medical Internet Research* 20 (6), e10486.

Lisenbee, P., Hallman, C. and Landry, D. (2015) Geocaching is catching students' attention in the classroom. *The Geography Teacher* 12 (1), 7–16.

Lund, K., Lochrie, M. and Coulton, P. (2010) Enabling emergent behaviour in location-based games. Proceedings of the 14th International Academic MindTrek Conference: Envisioning Future Media Environments (pp. 78–85). October. 6th–8th, 2010, Tampere, Finland.

Maas, J., Verheij, R.A., Groenewegen, P.P., De Vries, S. and Spreeuwenberg, P. (2006) Green space, urbanity, and health: How strong is the relation? *Journal of Epidemiology & Community Health* 60 (7), 587–592.

MacBride-Stewart, S., Gong, Y. and Antell, J. (2016) Exploring the interconnections between gender, health and nature. *Public Health* 141, 279–286.

Magaudda, P. (2012) What happens to materiality in digital virtual consumption? In M. Molesworth and J. Denegri Knott (eds) *Digital Virtual Consumption* (pp. 118–133). London: Routledge.

Maller, C., Townsend, M., Pryor, A., Brown, P. and St Leger, L. (2006) Healthy nature healthy people: 'Contact with nature' as an upstream health promotion intervention for populations. *Health Promotion International* 21 (1), 45–54.

Matthews, H., Taylor, M., Sherwood, K., Tucker, F. and Limb, M. (2000) Growing-up in the countryside: Children and the rural idyll. *Journal of Rural Studies* 16 (2), 141–153.

Mellou, E. (1994) Play theories: A contemporary review. *Early Child Development and Care* 102 (1), 91–100.

Moles, K. (2008) A walk in thirdspace: Place, methods and walking. *Sociological Research Online* 13 (4), 31–39.

NHS Digital (2015) Health survey for England 2015: Health, social care and lifestyles summary of key findings. Available at: http://content.digital.nhs.uk/catalogue/PUB22610/HSE2015-Sum-bklt.pdf (accessed 17 July 2017).

O'Brien, L. and Varley, P. (2012) Use of ethnographic approaches to the study of health experiences in relation to natural landscapes. *Perspectives in Public Health* 132 (6), 305–312.

O'Hara, K. (2008) Understanding geocaching motivations and practices (pp. 1177–1186). Proceedings of the SIGCHI Conference on Human Factors in Computing Systems CHI 2008, April 5–10, Florence, Italy. https://doi.org/10.1145/1357054.1357239

Oh, Y. and Yang, S. (2010) Defining exergames & exergaming. *Proceedings of Meaningful Play* 2010, 21–23.

Peng, W., Lin, J-H. and Crouse, J. (2011) Is playing exergames really exercising? A meta-analysis of energy expenditure in active video games. *Cyberpsychology, Behavior, and Social Networking* 14 (11), 681–688.

Ramsay, G., Dodds, R., Furtado, D., Mykhayletska, Y., Kirichenko, A. and Majedian, M. (2017) The barriers to millennials visiting rouge urban national park. *Sustainability* 9 (6), 904–916.

Richardson, E., Pearce, J., Mitchell, R., Day, P. and Kingham, S. (2010) The association between green space and cause-specific mortality in urban New Zealand: An ecological analysis of green space utility. *BMC Public Health* 10 (1), 240.

Richmond, D., Sibthorp, J., Gookin, J., Annarella, S. and Ferri, S. (2018) Complementing classroom learning through outdoor adventure education: Out-of-school-time experiences that make a difference. *Journal of Adventure Education and Outdoor Learning* 18 (1), 36–52.

Robinson, S. and Hardcastle, S. (2015) Exploring the attitudes towards and experiences of geocaching amongst families in the community. *International Journal of Environmental Health Research* 26 (2), 187–197. DOI: 10.1080/09603123.2015.1061116.

Romagosa, F., Eagles, P.F. and Lemieux, C.J. (2015) From the inside out to the outside in: Exploring the role of parks and protected areas as providers of human health and well-being. *Journal of Outdoor Recreation and Tourism* 10, 70–77.

Rosier, J.L., Yu, D. and Hung-Chih, A. (2011) Exploring the effects of geocaching on understanding natural resources and history. *Proceedings of the Northeastern Recreation Research Symposium, 2011.*

Ross, N.J., Renold, E., Holland, S. and Hillman, A. (2009) Moving stories: Using mobile methods to explore the everyday lives of young people in public care. *Qualitative Research* 9 (5), 605–623.

Schaal, S. and Lude, A. (2015) Using mobile devices in environmental education and education for sustainable development – comparing theory and practice in a nationwide survey. *Sustainability* 7 (8), 10153–10170.

Schlatter, B.E. and Hurd, A.R. (2005) Geocaching: 21st-century hide-and-seek. *Journal of Physical Education, Recreation & Dance* 76 (7), 28–32.

Schneider, J. and Jadczaková, V. (2016) Mutual impacts of geocaching and natural environment. *Acta Universitatis Agriculturae et Silviculturae Mendelianae Brunensis* 64 (5), 1739–1748.

Shanahan, D.F., Lin, B.B., Gaston, K.J., Bush, R. and Fuller, R.A. (2014) Socio-economic inequalities in access to nature on public and private lands: A case study from Brisbane, Australia. *Landscape and Urban Planning* 130, 14–23.

Sheller, M. and Urry, J. (2004) *Tourism Mobilities: Places To Play, Places in Play.* London: Routledge.

Shores, K.A., Scott, D. and Floyd, M.F. (2007) Constraints to outdoor recreation: A multiple hierarchy stratification perspective. *Leisure Sciences* 29 (3), 227–246.

Skinner, H., Sarpong, D. and White, G.R. (2018) Meeting the needs of the Millennials and Generation Z: Gamification in tourism through geocaching. *Journal of Tourism Futures* 4 (4), 93–104.

Sports Wales (2014) Calls for action – poverty factsheet. Available at: http://www.sport.wales/research--policy/tools-and-resources/publications.aspx (accessed 8 December 2018).

Staiano, A.E. and Calvert, S.L. (2011) Exergames for physical education courses: Physical, social, and cognitive benefits. *Child Development Perspectives* 5 (2), 93–98.

Smith, T.A. and Dunkley, R. (2018) Technology-nonhuman-child assemblages: Reconceptualising rural childhood roaming. *Children's Geographies* 16 (3), 304–318.

Van den Berg, A.E., Jorgensen, A. and Wilson, E.R. (2014) Evaluating restoration in urban green spaces: Does setting type make a difference? *Landscape and Urban Planning* 127, 173–181.

Whatmore, S. (2006) Materialist returns: Practising cultural geography in and for a more-than-human world. *Cultural Geographies* 13 (4), 600–609.

Wood, C., Sandercock, G. and Barton, J. (2014) Interactions between physical activity and the environment to improve adolescent self–esteem: A randomised controlled trial. *International journal of Environment and Health* 7 (2), 144–155.

Zemko, M., Vitézová, Z. and Jakab, I. (2016) Geocaching as a means for modernization of educational process. European Conference on e-Learning (p. 709). Academic Conferences International Limited, October.

Zbuchea, A. (2018) The digital tourist. *Journal of Tourism Challenges and Trends* 11 (1), 71–98.

12 Gamification: Augmented Reality, Virtual Reality Games and Tourism Marketing Applications

Evrim Çeltek

Introduction

Augmented reality (AR) is a technology that adds 'flavour' to real-world environments by enriching them with virtual objects. AR utilises devices such as smartphones or tablets to expand 'reality'. They help gain information about users' immediate surroundings via a kind of geolocation and use this information to superimpose virtual objects into reality (Buhalis *et al.*, 2019).

Virtual reality (VR) is a computer technology that allows users not only to replicate an environment, either real or imagined, but also includes the simulation of the user in that environment, which is achieved by mainly appealing to the vision and hearing of users via a headset, like the Oculus Rift, used in the application. VR is an artificial, computer-generated experience mostly utilised in gaming.

Games help promote brands at different levels by either including the product in the game without giving it an active role or making the product an integral part of it. There are several opportunities for companies to include AR and VR games into promotional campaigns. With any game, there is the potential for promoting and branding because people get pleasure from games that use AR and VR applications since they connect the digital world with the real world. Businesses will take the advantage of this trend by making their own AR and VR games or by profiting from existing ones. Therefore, the main purpose of this chapter is to look at the potential of AR and VR games in tourism marketing. The chapter presents the profile of AR and VR game applications and their marketing advantages for the tourism industry. In this chapter, 16 AR and VR game applications are examined, namely; Pokémon GO, Tripventure, Explore, Ingress, REXplorer, Loquiz, Avastusrada, Harry Potter: Wizards Unite, City Explorer, Vacation Simulator, TheBlu VR, Guided Meditation VR, Nature Treks VR, National Geographic Explore VR, Ocean Rift and National Archaeological Museum.

Augmented Reality and Virtual Reality

AR is the process of using technology to add digital elements onto a live-streaming video or the user's view of the real world. Used through smartphones, the AR will benefit from GPS technology letting users identify their location to control the device orientation. This technology captures a current image and integrates the new information into the picture (Berryman, 2012; Craig, 2013; Höllerer & Feiner, 2004). The AR system aims at improving the user's perception of and interaction with the real world by adding 3D virtual objects into the real world. In this way, they will seem to coexist as if they are real (Azuma, 2017; Azuma *et al.*, 2001).

AR allows the users to create 3D images or videos over any surface ranging from printed material such as advertisements, brochures or magazines to coffee cups by merely pointing the camera of their smartphones or tablets at certain objects (Azuma, 1997; Linaza *et al.*, 2012). AR also acts as a facilitator in the tourism business by offering a new world by integrating the digital world into the real one. As it is far too different from conventional advertising methods, it attracts tech-lovers (Craig, 2013).

AR- and VR-type technologies have already been used by companies in producing various types of tourism products to attract tourists. AR and VR technologies provide a variety of advantages for tourism marketing (see Table 12.1). To illustrate, many hotels (e.g. www.showhotel.com) and destinations (e.g. www.virtualgettysburg.com) offering 'virtual tours' are found on the internet. These 'virtual tours' often consist of panoramic photographs that do not allow any free navigation, which means that they are artificial VR. Despite this, they can still capture the attention of VR-type technologies (Guttentag, 2010). A Second Life (SL)-type virtual world has been created by a VR travel community called 'Itchy Feet' (http://www.itchy-feet.org) which allows tourists to search for travel information, communicate with other tourists and make travel purchases (Cheong, 1995).

Tourists may find AR practical as historical sites and museum exhibitions are full of detailed information that is not always what tourists are looking for. Therefore, museums, heritage sites, cities and tourist professionals may be expected to prepare layered information in accordance with the needs of people who might benefit from AR. In other words, considering the needs of tourists, regarding various factors including age, occupation, information level and so on will provide them with the

Table 12.1 Advantages and challenges of AR and VR for marketing

	Augmented reality	Virtual reality
Advantages	• Its fun feature makes brands memorable. • Customised communication because the user is put in charge. • Space- and time-free, unlike traditional media. • Lets customers share personalised content with others. • More cost-efficient development unlike traditional media. • Possible access to location-based information near the users. • Possible access to numerous up-to-date content. • Flexibility in the delivery of audio and visual media. • Increased interactivity with the integration of web-based map services. • Increased brand awareness. • Brand reputation. • Geo-targeting. • Interactive marketing. • Personalisation, viral marketing, socialisation, emotional connection and repeat engagement.	• Cost-effective advertising. • Boosting interest in visits to museums, which in return serves as a marketing strategy for tourist destinations. • Allows tourists to experience previews of destinations and their respective attractions and facilities. • VR systems also can also be used for direct marketing. • Allows trials before the purchase. • More engaged customers • Increased brand awareness. • Brand reputation.
Challenges	• Technologically complex to operate. • No standards yet. • The negative impact of outdoor surroundings, such as surrounding high buildings, on GPS systems, which is likely to happen indoors as well. • Privacy concerns. • Ethical concerns and other issues. • User consent required to participate. • Users' limited knowledge on how to operate.	• User consent is required to participate. • The health effects are still unknown. • Users' limited knowledge on how to operate.
Applications	• Starbucks (advertisement campaign). • The St Petersburg & Clearwater Convention and Visitors Bureau (interactive travel brochure). • Nova Scotia Tourism (guiding). • Jersey Tourism Agent, Stella Artois Le Bar Guide (advertisement and guiding). • Goggles and Layar (advertisement and guiding). • Hong Kong Tourism Board (guiding).	• Virtual tours of a business environment. • Training of new employees. • A 360° view of any product. • A virtual museum, gaming, virtual theme parks, interactive exhibitions, etc.
Types	• Marker based augmented reality. • Markerless augmented reality. • Projection-based augmented reality. • Recognition-based augmented reality. • Superimposition-based augmented reality.	• Non-immersive virtual reality. • Semi-immersive virtual reality. • Fully immersive virtual reality.

(*Continued*)

Table 12.1 (*Continued*)

	Augmented reality	Virtual reality
Tools	• Personal computers with webcams. • Kiosks, digital signage and window displays. • Smart phones and tablets. • AR glasses and head mounted displays.	• A computer with high processing capacity. • Headset: the head set also consists of three main components. These include sensors (magnetometers, accelerometers, gyroscopes), lenses and a display screen. • Input devices (such as joysticks, gloves, pedals and motion platforms such as Virtuix Omni).
References	Çeltek (2015); Chou and ChanLin (2012); Chung *et al.* (2015); Cranmer (2017); Fino *et al.* (2013); Fritz *et al.* (2005); Han *et al.* (2014); He *et al.* (2018); Höllerer and Feiner (2004); Kennedy-Eden and Gretzel (2012); Kounavis *et al.* (2012); Kourouthanassis *et al.* (2015); Lashkari *et al.* (2010); Lee (2012); Marimon *et al.* (2018); Selvam *et al.* (2016); Sing and Pandey (2014); TomDieck and Jung (2017); Wang *et al.* (2013); Wasko (2013); Weber (2014); Xu *et al.* (2017); Yovcheva *et al.* (2013).	Bonetti *et al.* (2018); Castro *et al.* (2017); Gibson and O'Rawe (2018); Guttentag (2010); Huang *et al.* (2016); Moorhouse *et al.* (2018); Tavakoli and Mura (2015); Tromp (2017); Tussyadiah *et al.* (2018); Williams and Hobson (1995); Xu *et al.* (2016, 2017).

maximum enjoyment of visiting a place and will help them gain a memorable experience (Berryman, 2012; Carmigniani *et al.*, 2011).

A variety of mobile AR applications (see Figure 12.1) serve tourists in many aspects including seeking, sharing or exchanging information and useful advice in addition to comments on a location or destination with a large network. Hence, by sharing experiences among many users, in this

Figure 12.1 AR applications from tourism: (1) Dubai Postcards (advertisement); (2) Starbucks Cup Magic for Valentine's Day (advertisement, gamification); (3) DiscoverHongHong AR (guiding)

Source: (1) https://www.youtube.com/watch?v=650jnMilsf8, (2020); (2) https://www.youtube.com/watch?v=8nvqOzj q10w, (2020); (3) https://www.youtube.com/watch?v=Iw15uXQ4puY, (2020) 1. Use QR codes to access AR apps.

Figure 12.2 VR applications from tourism (1) TheBlu VR; (2) National Geographic Explore VR

Source: (1) https://www.youtube.com/watch?v=A8XIYjNoovo, (2020); (2) https://www.youtube.com/watch?v= D1VILv6Db7I(2020). Use QR codes to access VR apps.

case, tourists, may be attracted through developing connectivity (Russell, 2012). Moreover, mobile AR applications can be useful in terms of providing tourists with the information they need only in accordance with their needs by adjusting the settings of the applications rather than an overdose of unnecessary information (Azuma, 2015).

VR application makes it possible to create 3D, interactive and computer-generated environments, which might have the replications of real or imaginary worlds as a way to provide information to users (see Figure 12.2). The VR technology has become more widely known as a result of a combination of different disciplines such as psychology, computer graphics, cybernetics, real-time and distributed systems, data-based design, multimedia, robotics and telepresence (Gurau, 2007). Virtual environment (VE), a 3D environment generated by computers, makes it possible for users to interact and navigate among themselves in this simulated area using at least one of their senses, such as hearing or seeing, and this is called virtual reality. The word 'interact' is about mobility in the VE, that is, the ability to move and replace objects in this computer-generated environment; 'navigate' refers to the ability to move around there (Guttentag, 2010). VR is considered to be a triggering tool to increase marketing in the tourism industry (Williams & Hobson, 1995). VR might be utilised in many phases such as planning, managing and promoting a destination in tourism (Cheong, 1995). Apart from tourism industry, VR is widely used in the fields of media, films, medicine, education, recreation and video games.

Augmented Reality and Virtual Reality Games

Video or computer games are designed mainly for entertainment, which involves interaction with computer-generated characters or tokens in a computer-generated environment (Eyles & Eglin, 2008). Such games for single players involve a series of difficulties that the players need to overcome one after another to win the reward at the end. In a multiplayer game, there are still difficulties to overcome, yet players are able to overcome them with the help of other players to gain the reward (Eyles & Eglin, 2008). Generally, computer games are played via devices such as computers and smartphones, and the virtual game environment is displayed in the confined area on screens. This puts the players in a different world, disconnecting them from the real world around them. Pervasive games, which have recently been released and attracted academic attention, offer a game environment that extend out into the real world (Sekhavat, 2016).

Business models can be varied with potential digital games based on AR serves as guiding, advertising and gamification. AR will increase the opportunity to purchase with 360° product views, which help customers virtually visualise the product in their homes. AR also enables customers to experience the product before they buy it, whether online or from a retail store. They will even be able to try on a piece of clothing virtually and see how it looks on them by using AR. A recent analysis in advertising indicates that demonstrating 3D models of a product to customers shopping online helps them to find extra information about products, develop a higher brand attitude, and accordingly, an increase in the intention to purchase. By clouding the line between real and computer-generated worlds, AR offers customers additional information realised by seeing virtual objects as if they are in the real-world on the screen (Sekhavat, 2016).

The implementation of a 3D artificial atmosphere in computer games is called VR gaming. VR atmospheres are produced with VR software and presented in such a realistic atmosphere that users feel as if they are in a real-world environment. A VR game may include a 3D image that can be explored on a computer device interactively by using the keys, mouse or touch screen, in its simplest form. Numerous advanced and immersive examples include VR headsets, wrap-around screens and VR rooms with sensory components such as wearable computers, scents and haptic devices for tactile feedback. VR gaming has been a dream for some people since the beginning of videogames and particularly 3D games. VR games used projector rooms or multiple screens.

In the control of VR games, tools such as a standard keyboard and a mouse, motion capture and game controllers are utilised. For more complex VR rooms, other tools may be used, such as treadmill floors, in order to give freedom to the users to move and feel in the VE. In

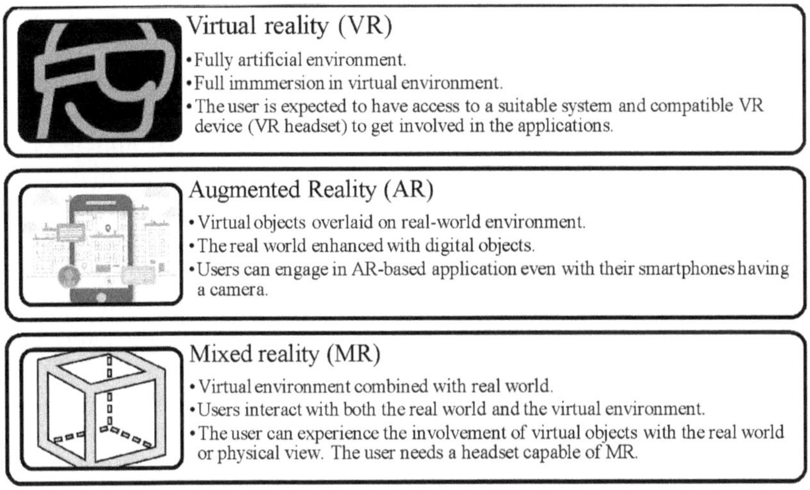

Figure 12.3 Differences between virtual reality, augmented reality and mixed reality
Source: Adapted from Tokareva (2018)

some VR gaming set-ups, the users are confined to the limits of the VE; that is to say, their movements are allowed within a certain area, but not outside of it. Yet, with the advancements in technology, VR headsets, without the other tools utilised, provide a greater experience to the users with less negative effects on them such as nausea. Different brands of VR headsets such as HTC Vive, Oculus Rift, Samsung VR, PlayStation VR, Google Cardboard and Microsoft Hololens are already available on the market. Most of these come with motion sensing controllers and trackers that work in accord with the headsets. Some others also have room-scale VR capabilities, and the others are used in sitting or stationary positions. In addition to the entertainment function, VR games also serve as tools for training and virtual therapy (Rouse, 2016a).

Location aware games utilising AR and mixed reality technologies (see Figure 12.3) primarily aim to alter the users' perception of reality; for example, in terms of their presence in a place either physically, socially or temporarily. As McCall *et al.* (2011) define it, AR gaming integrates audio and visual content into the real-time environment of the users. Indeed, virtual, location-related and contextual information are embedded into a physical site in AR game environments. To enable the game participants to access virtual information, computing devices such as mobile phones and hand-held computers are used in VR games because they are mobile and omnipresent. As stated by Schrier (2006), in VR games, game participants are simulated in a VE, which requires confined areas; yet AR games embed physical world, that is, the existing one present in real time, with digital information, communication abilities, networking and many other properties. Similarly, to play VR games, special VR headsets are required; however, except for a few, AR games are played on smartphones, tablets and other mobile gaming systems (Rouse, 2016b). In an AR game, a pre-created environment is superimposed over the present environment of the users, which might be something simple like a board game simulated on a table surface, or more complex, like an environment built from whatever is present in the users' real environment (Azuma, 2016). For instance, in such an AR game, there may be in-game characters climbing from coffee tables to sofas on virtual bridges. The demand is high for such AR environments because once the users explore one completely, the game loses its attraction. Yet, environment creation is a time-consuming task. Pokémon GO utilises a variety of tools such as the camera of smartphones, gyroscope, clock and GPS, in order to create a location-based, AR environment, which is considered a breakthrough AR app for gaming. In this game, the screen displays a map of the users' environment in the present. The presence of a Pokémon is indicated through a rustle of grass. When the users tap on the touchscreen, it captures the Pokémon. In AR mode, the Pokémon appears within the users' environment (Rouse, 2016b).

Tourist AR games may be designed for single or multiple players, for example, family members or friends or other travellers, and offer several activities such as visits to museums or photo shoots. This makes the travel experience unique for the tourists, hence the increased quality of the tour. In addition, such AR games help raise the awareness of tourists about travel destinations and tourist attractions in an entertaining way. During the game, the players are required to fulfil some tasks such as buying tickets for transportation or for places for interest, either in their immediate surroundings or in remote areas. This may actually result in longer tours as tourists may need to stay longer to fulfil these tasks, which means extended stays at hotels and more reservations at restaurants. City branding is also promoted through this; that is to say, they make a destination more and more popular for potential tourists.

Using AR and VR games in marketing applications in the tourism industry has several advantages for businesses and customers. These advantages are explained below:

- Immersive and interactive user experience: personalised headsets and VR solutions make it possible for VR enthusiasts to experience many things ranging from travelling to their dream destinations to certain emotions. The possibilities offered by VR means that it is as if these experiences are happening both in front of them and to them (Hassan & Rahimi, 2016; LaValle, 2019). As for AR, similar to VR, users may add a personal touch to the 'augmented' content, either online or published, performed with the help of custom-made smartphones and the markers that offer more information (Cranmer, 2017; Weber, 2016). For instance, in destination treasure hunt games, gamers learn as they screen the destination history with AR apps; they can also gain additional information regarding culture, tourism facilities and things to do.

- Personalised content: the AR treasure hunt games may be played according to the tourist's choice of game place (Weber, 2016).
- Help with customer shopping for choices and decisions: a destination or hotel VR game can influence the customer's buying decisions (LaValle, 2019). Cities can use VR games to introduce their sightseeing spots, as well as local food and different attractions. Experiencing a world of VR might stimulate viewers to travel to its authentic world. Putting on a headset and teleporting to an exotic destination while learning about the culture and places of interest could help sell travel packages for agencies and hotels.
- Users have exciting experiences: VR technology appeals to the senses of the users, that is, they feel as if they are in the real location, including the sounds and objects. This motivates users to use VR more (Tussyadiah *et al.*, 2018). In particular, people with disabilities may appreciate this more because they might be able to explore the real world through VR. Games utilising VR allow users to experience the whole surroundings, which creates interactive viewing expertise for their users.
- Detailed views provided: as LaValle (2019) states, VR games provide a full view of a place which can make some tourist destinations more interesting and even more accessible. In other words, users may have a clear idea of places that they would like to explore, including the tourist attractions they are interested in. This actually helps them plan their trip easily.
- Effective communication: as Weber (2016) states, the effective communication that VR and AR games provide is its top advantage. Users can communicate with each other while playing games, which is a new way of experiencing of communication, and hence probably more enjoyable.
- Builds brand loyalty and collects information about customers: games are considered essential tools to building brand loyalty and collecting information about current and prospective customers as they have built in advertisements. This promotes branding and product awareness and gathers information about the customers (Gura *et al.*, 2018). As for the branding of tourist destinations, VR games are important and innovative tools because they include a mixture of brand elements to create a positive image to reach the target market (Santos *et al.*, 2007).

Using AR and VR games in marketing applications in the tourism has challenges. These challenges are explained below:

- High cost: one of the widely accepted drawbacks of VR games is that they are not cost efficient, that is, they are not affordable for everyone. It is expensive and people who cannot afford them will be excluded from using this technology.
- Users become addicted to the virtual world: the users become addicted to the virtual world and spend a great deal of time navigating the non-virtual environment. This addiction may cause several health problems such as eye disorders, loss of reality and aching joints.
- Technology is still experimental: although VR technology is employed in varied fields, it continues to be experimental. The use of it has not expanded to many areas of life because it is not yet fully developed. This affects the ratio of its acceptance among people; that is to say, it does not receive a great deal of acceptance.

VR games simulate real-life experiences and users feel more engaged, which is a part of the gaming process. 3D glasses, game consoles and gloves provide a great experience in VR games, which makes them more engaging. AR games, on the other hand, utilise computer-generated objects that are implemented into real-life situations, hence there is more interactivity because users can relate to the computer-generated characters. AR and VR, to some extent, use the same space; the former using virtual characters in real world setting and the latter using computer-generated characters and surroundings to simulate a real-life setting. AR and VR differ in the way they are used. For a VR experience, VR headsets and 3D glasses are required, yet for an AR experience, applications installed on smartphones or other similar mobile devices are sufficient, which makes it easy to use. In addition to this, a VE is needed for AR games as in mixed reality (MR), unlike VR games, because they are self-sufficient with their own virtual world (Augray, 2018).

Methodology

AR and VR game technologies are likely to evolve and flourish, with new possibilities in the tourism sector. Yet, there exist many AR and VR games in the tourism sector regardless of their anticipated evolution. The analysis of these games may give insights into how the tourism sector might benefit from them, the opportunities they offer and what impact they will have on the sector. This study aims to shed light on the benefits of AR and VR game applications for tourism researchers and professionals regarding marketing opportunities.

The data collection process follows closely that of Yung and Khoo Lattimore (2017), who employed the method for their systematic quantitative literature review of VR and AR research in tourism. The five-step review protocol consists of: (1) determining review aims and formulating research questions; (2) identifying search terms and selection criteria; (3) searching for games before refining exclusion and inclusion criteria; (4) appraising game quality and relevance; and (5) synthesising and reporting findings.

The games in the study were limited to tourism. Others were excluded because this research aims to determine the VR and AR games in tourism. Given this study's review aims, the following search keywords were used: 'augmented reality game', 'AR game', 'virtual reality game', 'VR game', 'augmented reality game examples', 'virtual reality game examples', 'best augmented reality game examples', 'best virtual reality game examples' and 'augmented reality and virtual reality game examples in tourism' were used to search for relevant examples. After completing the game search, the games were screened for inclusion and exclusion criteria.

The data were collected via the internet from 1 January 2018 to 10 April 2020. First, the AR and VR game examples were collected by using an internet webpage search. First of all, to find the examples, keywords such as 'augmented reality game' were used in a Google search. Then, the results, including webpages, blogs, articles and corporate websites that offer AR game and VR game software or marketing, were examined individually. Viewing all these webpages, blogs and articles led to other examples. There appeared to be numerous AR and VR applications used for guiding and advertising in the tourism industry, yet few AR and VR games because they are at the development stage, hence new for the tourism industry. As a result of exploring the examples available on the internet, 16 AR and VR game applications were identified, namely Pokémon GO, Tripventure, Explore, Ingress, REXplorer, Loquiz, Avastusrada, Harry Potter: Wizards Unite, City Explorer, Vacation Simulator, TheBlu VR, Guided Meditation VR, Nature Treks VR, National Geographic Explore VR, Ocean Rift, National Archaeological Museum.

Results

Pokémon GO

Pokémon GO (see Figure 12.4) was released on 6 July 2016, and destinations around the globe have been incorporating the AR game into their promoting efforts ever since. The game was installed over 100 million times less than a month after its release (Cranmer, 2017; Koikawa, 2016). The game asks players to explore common and little-known landmarks alike and allows destinations to increase the period of time locals and visitors spend at businesses, including hotels. The game permits players to go looking for Pokémon, fictional animals derived from the internationally recognised Pokémon franchise. The game uses geolocation to create AR gaming scenarios for players (Wagner-Greene *et al.*, 2017; Weber, 2016) and requires players to move (walk, bike, drive, etc.) to

Figure 12.4 Pokémon GO
Source: Goldfarb (2020). Use QR code to access image.

capture free-roaming Pokémon creatures and train them to fight different other creatures; this includes walking designated distances (2 km, 5 km or 10 km) to 'hatch' Pokémon eggs. The game also allows the players to look for and catch over 100 species of Pokémon as they move around their surroundings (Koikawa, 2016). A Hong Kong hotel also gives fans incentives and prizes to keep them coming back. Beginners who have caught their first 15 Pokémon around the city are entitled to a 15 % discount at the hotel restaurant as well as a 30 % discount at the hotel bar. In addition, more skilled trainers who have caught their first 20–50 Pokémon can receive a complimentary one-night stay once they book a night at Ovolo, and the hotel will also give a free upgrade and a welcome Pokémon cocktail (Newgenapps, 2017). In the three-month period after its release in July 2016, the game was downloaded over 550 million times and has created over US$470 million in sales (Niantic and Nintendo 2016; Wagner-Greene *et al.*, 2017; Weber, 2016). Though the peak of the game happened throughout its first month, Pokémon GO is still downloaded over 700,000 times every day worldwide (Wagner-Greene *et al.*, 2017).

Game type

Location-based AR game. (It is currently active for playing.)

Target group

Tourists who travel in any city.

Marketing benefit

Promotes city and destinations, historical places and buildings. If done properly, it can create real connections between business, its products and consumers. If companies aim to maximise customer loyalty, customers need to be rewarded to keep investing in their product. In Pokémon GO, incentives and bonuses offered when players go one level up, take on gyms, catch new Pokémon, walk to find a rare Pokémon or win a battle keep the players active even in less active parts of the game. These rewards ensure the continued investment and keep the players engaged in the game, sometimes at the expense of their productivity. Pokémon GO is a great way to lure new customers into the business neighbourhood. Niantic, the game-maker, makes it possible for businesses to be featured in the game as new PokeStops and Pokegyms. This helps them attract visitors in search of Pokémons using a 'lure module'. For instance, the British Museum have PokeStops in the museum's public areas and hopes that gamers looking for Pokémon's may also turn into visitors purchasing museum tickets.

Marketing applications

The game asks players to explore common and little-known landmarks alike and provides destinations to increase the period of time locals and visitors spend at businesses, including hotels.

Marketing communication

It is a multiplayer and single player game; that is to say, it can be played alone, or it can be played by groups or friends. Multiplayer games have advantages in interactive marketing, marketing communication with the customer, customer loyalty and WOM affect. Users can communicate with each other and therefore enjoy a conversation. AR games may also be a strong tool to build brand loyalty and capture crucial information regarding existing and potential customers. They contain unique interactive games with a suite of advertising creatively built in. By combining games and ads, it is likely to enhance branding, boost product awareness and collect detailed knowledge about participants.

Problems with the game

The Pokémon GO game has led to several high-profile incidents, including the death of players. There has even been a Pokémon GO Death Tracker set up for the purpose of providing more information about people who have died while engaging with the AR app. To prevent all these, regulatory guidance is needed in the field. The game can cause digital fatigue as well as addiction. The game requires approval or participation from the user.

Tripventure

Tripventure games and guides (see Figure 12.5) are location-based games that can be played in many cities in Europe. They allow tourists to experience virtual stories in real life via AR. Examples include

Figure 12.5 Tripventure AR game iPhone screenshots
Source: http://www.148apps.com (2018). Use QR code to access image.

Inspector Tripton (Find the murderer!), Secret City (What happened to an old friend?), Fall of the Wall (How did the Berlin Wall come down?) and Rocco Ratcha (Collect all the pizza slices), which presents a combination of reality and fiction. While exploring the actual city, tourists interact with virtual people, solve mysteries and find hidden items to reach the answers to the questions given (Appdedice.com, 2018).

Game type

Location-based AR game. (It is currently active for playing.)

Target group

Tourists who travel in European cities.

Marketing benefit

Promotes European cities with virtual stories. The Tripventure app presents a variety of tours and games for players who are looking for adventurous location-based gaming or want to discover a city differently. Story-based gameplay meets AR. Tripventure reveals the city's secrets and turns the city into a playing field. Using the smartphone, players see more than ever before, meet characters, solve riddles or find virtual objects. A museum becomes a crime scene, phone booths turn into a meeting place and landmarks into sites of action.

Marketing applications

Tourist experiences virtual stories in the real world. The game combines reality with fictional stories.

Marketing communication

It is a multiplayer and single player game. It can be played alone, in a groups or with friends. Gamers who play Tripventure in European cities can communicate in an interactive way with other gamers who play. Therefore, the game has advantages in interactive marketing, marketing communication with the customer, customer loyalty and WOM affect. Tripventure can be a strong tool to build city brand loyalty and capture crucial information regarding existing and potential tourists who travel to European cities. Gamers travel to European cities in an entertaining way. This game can be defined as an opportunity where both institutions and destinations can differentiate themselves in the eyes of incoming tourists. It is particularly important in terms of providing a competitive advantage in destination marketing.

Problems with the game

The game can cause accidents when the player interacts with the virtual characters in the game, which is a safety concern when using AR in an outdoor setting. It can cause digital fatigue and addiction. The addiction may cause varied health problems for players. Gamers may find AR and VR games interesting, but the everyday consumer sees no need to purchase these games. In the game, there are crime and murder stories that can be are harmful to children, thus an age restriction should apply.

Explore

Explore is the name of a gaming and guiding system used in museums and exhibition environments. This game aims at creating a location-sensitive, interactive, guiding, infotainment and entertaining system for museums. It is accompanied with new media offering information to the visitors. It provides guests with virtual tours, information and some games (trivia). The system is based on Bluetooth technology and the guests' usage of their mobile phones as terminal devices. The system can be adjusted to more than one museum and exhibitions simultaneously, and it can be built on different data management platforms. Socially, it aims at increasing the attendance of people to museums and cultural exhibitions by making them attractive. It has two important target audiences; the first is children who are occupied by various entertainment services; the system provides them an alternative way of entertainment. The second target audience is the foreign guests of the exhibition who are served with multilingual equipment by the system. This system works by integration with the network of European museums through the internet and in this way, it highlights the common features among European countries as well as individuals visiting the exhibition at the same time. The project started in February 2004 and was finalised in October 2005 (Lopatina, 2005: 22). The Explore system provides location-dependent games, such as trivia. When the visitor chooses a tour, they can download a group of questions based on the preferences of visitors at the different places in the museum. The results of the trivia are kept in the portal session of the museum. The visitor will be able to read the results and compare their results with the ones of different visitors either in the middle or at the end of the session. Museum employees are the creators of all the questions during the exhibition design step (http://explore.do.isst.fraunhofer.de./technology.html).

Game type

Location-based AR game. (It is no longer active for playing.)

Target group

Tourist as visitors of museum and exhibition environments.

Marketing benefit

Makes museum and exhibition environments attractive to tourists. It is a game-based application for museum and exhibition environments.

Marketing applications

It provides visitors with the experience of virtual tours, helps them receive information about the museum and play games (trivia). Bluetooth technology is the main part of the setting of the system. It is accompanied by a terminal device which is the visitors' own mobile phone. This system is integrated into the network of European museums through the internet and thus it highlights the common characteristics among European countries as well as individuals visiting exhibition at the same time. When the visitor chooses a tour, they can download a group of questions based on the preferences of visitors at the different places in the museum.

Marketing communication

It can serve both as a multiplayer and single player game; in other words, it can either be played alone or with groups or friends. Multiplayer games are advantageous in terms of interactive marketing, marketing communication with the customer, customer loyalty and WOM affect. Players experience interacting with others while playing the game, and meanwhile, it increases the satisfaction of tourists with the gamification of museum visiting. Tourists experience an interactive museum visit which makes them likely to revisit the museum. Passing an enjoyable time in the museum with gameplay, the tourists might share this with their acquaintances, on social media, and in this way, they can help increase the demand for the museum. These are all significant and necessary benefits for marketing communication.

Problems with the game

The users may become addicted to the virtual world and spend a lot of time navigating the non-virtual environment. This addiction may cause varied health problems for players. Also, as the application is played with Bluetooth on, it may decrease the security of the player's phone. The phone may become vulnerable to malware. Finally, visitors who interact with the game in the museum might be noisy and disturb others who are not in the game.

Ingress

Ingress is a story-driven, real-world AR game. Players produce the Portals. To recommend a new Portal, one needs to take a photograph of the desired place, monument, building or object, provide it with a name and short description, and submit it via the official game website, ingress.com. There are certain criteria for a Portal; for instance, it has to be distinctive, unique and accessible, and may not include private residential properties, primary/secondary schools, hospitals and fire/police stations. The Portals can be found at the monuments, notable buildings and different places of cultural significance or interest. The game additionally rewards exploration, thus players are incentivised to try new restaurants, visit museums or simply walk. Players are also given in-game badges when they hack or capture Portals they have never visited before. Also, the games offer 'missions', completion of which may win players special badges, and these missions require interaction between multiple Portals in a given place. For example, if players visit the Smithsonian's National Air and Space Museum in Washington, DC, they will have various missions to be completed while exploring the exhibitions there. Some of these missions might be simple ones, yet some others might require strategic planning (Lifehacker.com, 2018).

Game type

Location-based AR game. (It is currently active for playing.)

Target group

Tourists who travel to cities.

Marketing benefit

Ingress not only urges players to walk, but also lets them explore the surroundings in a completely new way, discovering interesting buildings and objects around them that they have not been aware of, or might not have paid attention to otherwise.

Marketing applications

Players produce Portals. To recommend a new Portal, a player needs to take a photograph of the desired place, monument, building or object, provides it a name and short description. The game additionally rewards exploration, thus the player has an incentive to try new restaurants, visit museums or simply just go for a walk.

Marketing communication

It is a multiplayer and single player game, that is, it can be played alone, or it can be played by groups or friends. Multiplayer games have advantages in interactive marketing, marketing communication with the customer, customer loyalty and WOM affect. Tourists who are satisfied with their experiences in the game will want to repeat it and recommend it to their friends. This will provide a competitive advantage with word of mouth marketing as well as the region's own marketing strategies.

Problems with the game

While playing the game, the phone's charge may run out or internet connection may be lost as there is likely to be no or low-quality internet in some places. Also, players may become addicted, which may later cause various health problems. Lastly, the game can cause traffic accidents, although safety is the most important concern while using AR in outdoor settings.

REXplorer

REXplorer is a mobile 'spell-casting' game aimed at tourists visiting Regensburg in Germany, which is inspired by the idea that visitors go sightseeing and learn about the history of Regensburg, a historical city, while walking through it via MR storytelling techniques (Chen, 2008). REXplorer has participants complete missions located in different parts of the city, but it also encourages them to have virtual characters, once inhabitants of the city, and tells them stories. Participants are also given tasks, such as delivering a letter from a character to another one living in different part of the city. Altogether, they help participants learn more about the city they are exploring because such tasks require them to walk around the city, therefore making it more fun while at the same time acquiring knowledge and familiarity. As Azuma (2015) states, some participants find it as a way to integrate real life into a historical tour, which makes it interesting; otherwise, they would find the tour boring and dull.

The player comes across some ghosts (historical figures) who have relationships with significant buildings in an urban setting. A new mobile interaction mechanism called 'casting a spell' (making a gesture by waving a mobile phone in the air) helps the player to awaken and talk to the ghost to continue the game. The game aims to tell guests about the history in an amusing manner (Chen, 2008). REXplorer uses mobile social software concepts to strengthen the game and increase the tourist's satisfaction with the experience. The game is based on the idea that the historical buildings of Regensburg (Weber, 2016) have magical powers, secrets and treasures locked inside them. All these things need to be released with the proper magic spell. Magic spells are scattered around and by using a mobile phone in a plastic shell as a wand equipped with custom software that uses gestural recognition of the motion of the video from its camera to recognise certain basic movement shapes, and then calling 'ghosts' to talk about the city's history. Voices from magic creatures stuck inside the buildings are heard by players through their magic wand (the loudspeaker on the smartphone). The creatures will be unhappy and uncooperative if a player casts the spell incorrectly. However, if they manage to cast the spell appropriately, the spirits can open up to the participants and tell stories from the past including clues that lead them on their way to the master wizards. The game also has challenges, such as a spell-casting battle between participants to reach their goals (Kuniavsky, 2006).

Game type

Location-based AR game. (It is no longer active for playing.)

Target group

Tourists who visit the German city of Regensburg.

Marketing benefit

Promotes the city and its history to tourists. It is a city-focused game.

Marketing applications

The game platform blends location services, mobile phone photos and video blogging as well as phone camera-based interactions to create a fantasy world that brings the history and culture of Regensburg to life. The game helps tourists and visitors to explore the history of the UNESCO world heritage city of Regensburg, Germany, and magically interact with the city's monuments instead of just seeing them.

Marketing communication

It is a multiplayer game. Multiplayer games have advantages in interactive marketing, marketing communication with the customer, customer loyalty and WOM affect. REXplorer enables interaction between players and the environment of the city of Regensburg. Location-based mobile games are informative games that provide tourists with more detailed information about the place that they have travelled to. Further studies connected to these ideas and went beyond to enhance visitor learning in cultural heritage sites tell interactive stories of locations or advance the interaction between the visitor and touristic artefacts. Games have the power to create greater engagement with the destination through mediated and playful interactions (Weber, 2016).

Problems with the game

Addiction is again a problem, which may result in gamers becoming addicted to the virtual world. As the game is played via mobile phones, battery and internet problems apply here too. Also, accidents may happen despite the safety concern of AR games, especially when played in outdoor settings.

Loquiz

Scavenger hunt games have been around for a long time. They can be played for fun at parties, used for team building in firms or for tourists as a different kind of gamified sightseeing. Loquiz takes it to the next level by providing a mobile platform for making such games available on smartphones and tablets. The game moderator will produce a game by placing pins on a map in certain locations and add questions for every pin along with clues, hints, photos and even videos. Game participants in groups of three to five members must follow the map, move from pin to pin (in the order outlined by a moderator) and answer the questions. The app was developed in Estonia and has become quite popular in numerous travelling agencies and companies providing outdoor/recreation activities in Baltic States and Scandinavia, and also in other countries (Ivanova, 2016).

Game type

Location-based treasure hunt AR game. (It is currently active for playing, but it is not free.)

Target group

Tourists visiting Estonia, the Baltic States and Scandinavia.

Marketing benefit

Promotes the city and its history to tourists.

Marketing applications

A game moderator will produce a game by placing pins on a map in certain locations and add questions for every pin along with clues, hints, photos and videos.

Marketing communication

It is a multiplayer game. Multiplayer games have advantages in interactive marketing, marketing communication with the customer, customer loyalty and WOM affect. Scavenger hunt games can be played almost anywhere. The fact that nearly every aspect of each event can be matched to the tourism marketing communication requirements means that almost any request can be catered for. The flexibility in terms of the chosen clues, locations and prizes of the event means that the 'message' or objective of a scavenger hunt game event can be conveyed to tourists in an engaging fashion. The game has the advantage making a culture tour or city tour entertaining.

Problems with the game

Loquiz is a mobile game platform that enables the creation of adventure games and scavenger hunts. Games are created online and played on smart devices. One of the problems with Loquiz and in general with this kind of platform is that when you run scavenger hunts or any outdoor activity, you will need to deal with local authorities and seek permissions if your party size is substantial. Loquiz also charges for teams to enter the game, but there are no monthly fees.

Avastusrada

The project Avastusrada ('Discovery Trail' in Estonian), created by Tallinn University's (TLU) Institute of Ecology and Institute of informatics is not entirely an app; however, it is mobile (it works from a smartphone's browser) and it has similar concepts to the Loquiz game. Participants walk in the real world following predefined paths and reach target locations shown on the map in a certain order. Every location has its own directions (e.g. question or additional directions), which can be accessed only if the player is close enough to the target location. Most of the trails are placed in nature centres all over Estonia. The main purpose of the project is to educate people, particularly school students, regarding nature and the environment and how it is necessary to take care of it (Ivanova, 2016).

Game type

Location-based AR game. (It is currently active for playing.)

Target group

Tourists and students who live in Estonia.

Marketing benefit

Promotes the city and educates tourists and students about the nature and environment of Estonia.

Marketing applications

Participants walk in real world following predefined paths and reach target locations shown on the map in a certain order.

Marketing communication

It is a multiplayer game. Multiplayer games have advantages in interactive marketing, marketing communication with the customer, customer loyalty and WOM affect. Avastusrada players can communicate with each other and therefore enjoy a conversation, and they are educated about nature in an entertaining way. The game has edutainment advantages so it can build loyalty and capture crucial information regarding existing and potential customers in Estonia.

Problems with the game

The game needs GPS signals. While playing the game, the phone's charge may run out quickly. The internet may also not work everywhere. The game can cause traffic accidents, which is a safety concern while using AR in an outdoor setting. Finally, the playing area of the game is limited to Estonia and it has one language option (Estonian).

Harry Potter: Wizards Unite

Harry Potter: Wizards Unite is similar to Pokémon GO and other related games. Players travel in the real world, collect various things and fight different enemies. Players collectively deal with a disaster that affects wizards everywhere. Players will be able to establish themselves as an official member of the wizarding world and the Statute of Secrecy Task Force by personalising their selfie with Wizarding World-inspired lenses, frames, deformations, accessories, stickers and many more customisable features. The game uses geolocation to create AR gaming scenarios for players. While players explore the real world, they find a variety of wizarding locations on their map, including Inns, Greenhouses, Fortresses and Magical Traces where players can trigger specific encounters (Google Play, 2020).

Game type

Location-based AR game. (It is currently active for playing and free.)

Target group

Tourists or players who travel in any city.

Marketing benefit

Promotes cities and destinations, historical places and buildings. If done properly, it can create real connections between businesses, their products and consumers.

Marketing applications

In the game, players collect energy by visiting places called Greenhouses and Inns and help the magicians by using the collected energy in the game. Players look for virtual creatures in public spaces and businesses so it can be a cost-effective marketing tool to 'lure' people to businesses. Hotels and restaurants can turn players' energy gathering into a tool that contributes to the marketing of businesses.

Marketing communication

It is a multiplayer and single player game. Multiplayer games have advantages in interactive marketing and marketing communication with the customer, customer loyalty and WOM affect. Users can communicate with each other and therefore enjoy a conversation. AR games may be a strong tool to build brand loyalty and capture crucial information regarding existing and potential customers. They contain unique interactive games with a suite of advertising creatively built in. By combining games and ads, it has the potential to enhance branding, boost product awareness and collect detailed knowledge regarding participants.

Problems with the game

The game can cause digital fatigue and addiction to the game for players. The game also requires approval or participation from the user. Last but not least, the game cannot be downloaded and played in every country.

City VR

This game allows the player to walk on each water and land as the player looks around in an immersive virtual surrounding (see Figure 12.6). The application designed to be utilised in a VR viewer such as Google Cardboard. Game options include hundreds of streets, buildings, vehicles and different urban areas, each of which has its equivalent in real life, yet City VR offers more than just a 3D map. Players have many capabilities there; for example, they can travel through the streets like a giant, see things from up to down, zoom in and out, change the weather conditions

Figure 12.6 City VR game

Source: https://cityvr.com (2020). Use QR code to access image.

and take amazing snapshots. The available cities on the game include San Francisco, USA, where players stand on top of the Transamerica Pyramid, walk as a giant to the Coit Tower, sit in the AT&T Park, watch the sunset with the view of Golden Gate Bridge and teleport across the town and use time-lapse to replicate the rise again over the Bay Bridge. Chicago is also included, where players can visit the Millennium Park to observe the reflection of the town skyline on Cloud Gate. They are able to take 1450-foot leap from Lake Michigan to the SkyDeck at Willis Tower. New York, another upcoming city on the game, will allow the players to look down on Times Square and stroll on the Fifth Avenue (Google Play, 2018).

Game type

City simulation VR game. Needs VR headset, tracked motion controller and room scale play area. It is currently active for playing.

Target group

Players who want to discover San Francisco and Chicago in the virtual world.

Marketing benefit

Promotes San Francisco and Chicago. With the help of VR headsets, the player is fully immersed in the VE, which makes them believe they are inside the game when in fact they are still in their living room.

Marketing applications

Players discover hundreds of streets, buildings, vehicles and different urban areas. After seeing the cities in a virtual world, players may want to explore the city in the real world.

Marketing communication

It is a single player game. Brands or companies can advertise their products and promotions in the game. The City VR game can influence customer buying decisions. Cities can use the VR game to introduce their own sightseeing spots, as well as local food retailers and different attractions. Experiencing a world of VR might stimulate viewers to travel to its authentic world. VR games provide a full and detailed view of a place. For instance, VR games make tourism destinations more interesting and easier to marketing the destination. It offers a detailed view of a place wherever the player wishes to go. Therefore, viewers can organise their trip by seeing the locations of that place.

Problems with the game

The game requires a VR headset, a tracked motion controller and a room scale play area; however, in AR games, a smartphone is usually sufficient and is therefore more accessible and affordable to most players. For the VR game market to grow, and for VR to prove its value to consumers, prices need to adjust to consumer budgets. A VR headset is can be expensive for players. Unlike AR games, VR games require large volumes of data. The duration of the city explorer game is short. Players may want to travel to the destination or city in VR for a longer time, which will increase their data usage. Extended use of VR can lead to loss of spatial awareness, dizziness, disorientation and nausea. Also, using a VR headset can be harmful to the eyes or can cause other health problems. Finally, if someone spends an extended period in a VE, they may become addicted to it.

Vacation Simulator

In the Vacation Simulator game (see Figure 12.7) there are a dozen teleportation nodes, and each of them includes their own individual activities. There is a sandcastle building station, where the player can purchase items with sand dollars. Players receive mini quests to explore the cute, if not crowded, beach cove. They may build a huge sandcastle and save it on robot's floppy disk which is a character in the game. The game is similar to Job Simulator. The key distinction is that, rather than being directed to do a lot of menial job tasks, players can get to relax by the beach (Road to VR, 2018).

Game type

Vacation Simulation is a VR game. It needs a VR headset and tracked motion controller. It is currently active for playing and is not free.

Target group

Players who want to take a holiday in a virtual world.

Figure 12.7 Vacation simulator
Source: Road to VR (2018). Use QR code to access image.

Marketing benefit

Taking a holiday in a virtual world.

Marketing applications

The player receives mini quests to explore the beach. The game has potential to advertise on the virtual vacation.

Marketing communication

It is a single player game. Brands or companies can advertise their products and promotions in the game. A vacation VR game can influence the customer's buying decisions or allow the player to live a virtual holiday experience if they cannot go on holiday (e.g. people with disabilities). Experiencing a world through VR might stimulate viewers to travel to its authentic world. VR games provide a full and detailed view of a place. Since there is no possibility to try before they purchase tourism products, the virtual travel games provide the tourist with the opportunity to try before the actual product is purchased.

Problems with the game

For the VR game market to grow, and for VR to prove its value to consumers, prices need to adjust to consumer budgets. A VR headset can be expensive for players. Although the status is improving, currently the market has low VR penetration in households with limited broadband. There are less VR game examples and applications in the tourism industry and vacation simulation games are not being developed for marketing and branding the tourism businesses; they are being developed for fun. Users can become addicted to the virtual world and it can cause a loss of human connections for the player. Wearing the VR headset for long periods may result in fatigue and an unsettling feeling of enclosure.

TheBlu VR

TheBlu VR transports the player into a gorgeous VR experience that is not solely interesting and fun, but also educational. It allows the player to live alongside the creatures in the ocean, in their own habitat, untouched by humans, and not in a zoo or oceanarium. Also, the beautiful 3D graphics add to the excitement of the totally immersive underwater journey that TheBlu VR provides, whether the player is spending time with orcas in the Arctic, relaxing with dolphins close to the beach, or staring into the eyes of a huge blue whale. With varied underwater creatures swimming around the player's head, the VR headset seems like scuba goggles – except this one contains a calming voiceover guiding player each step of a way (Oculus, 2018).

Game type

Sea simulation VR application. It needs a VR headset, tracked motion controller and room scale play area. It is currently active for playing and not free.

Target group

Players who want to discover the sea in a virtual world.

Marketing benefit

Promotes the seas and oceans and gives information about the sea/ ocean creatures and habitat.

Marketing applications

Players live and dive in a virtual ocean or see and swim with dolphins, orcas and whales.

Marketing communication

It is not actually a game-based application but can be used as a VR game in tourism. VR games are much more entertaining and immersive than other marketing communication applications. Playing VR games gives the player the chance to step into a completely different world. This will make the entire tourism experience more realistic than advertisements. It gives a chance to experience new things at unknown destinations. Players can experience new destinations before making buying decisions. If they like the experience in the game, they might buy the real tourism tour. In other words, the biggest benefit of VR games will be the amazing perspective it provides on the overall tourism destinations, cities or tours. For instance, experiencing the ocean world with VR might stimulate viewers to travel to oceans in the real world.

Problems with the game

For the VR game market to grow, and for VR to prove its value to consumers, prices need to adjust to consumer budgets. For example, a VR headset may not be affordable for all players. Some VR headsets carry a high price tag, especially for proprietary closed-face designs. TheBlu game is not targeted marketing tourism; it was developed to provide information about the oceans. Also, wearing a VR headset for long periods produces fatigue, an unsettling feeling of enclosure, dry eyes and strain on eyes. Finally, as mentioned earlier, players need to purchase the game, that is, it is not free.

Guided Meditation VR and Nature Treks VR

Guided Meditation VR (free)

The application was developed for meditation. The virtual guide (in four languages) gives meditation directions to users in a VR world. There are more than 100 landscapes (landscapes from the real world) that can be travelled. Users can meditate and listen to music withing these relaxing and exotic places. If they do not want to meditate, they can visit these places and listen to music or the radio in the VR. It has 10, five or two-minute meditation options with music or radio.

Nature Treks VR (not free)

In the application, users discover tropical beaches, underwater oceans, nine breath-taking colour-themed environments, over 20 different animals (including elephants, giraffes, whales, dolphins), command the weather, take control of the night or shape the own world. Users also utilise the 'meditation Lotus' to help control their breathing and experience a world filled with the calming, soothing sounds of nature. This application developed for meditation and relaxation (Oculus, 2020).

Game type

Both meditation simulation VR applications. They need a VR headset, tracked motion controller and room scale play area. Both games are currently active for playing.

Target group

Users who want to meditate in relaxing and exotic places in the virtual world.

Marketing benefit

Promotes relaxing and exotic places in the world. With the help of VR headsets, users are fully immersed in the VE.

Marketing applications

Users explore relaxing and exotic places on the earth. After seeing the places in a virtual world, they might desire to explore these places in the real world.

Marketing communication

It is a single player game. Brands or companies can advertise their products and promotions in the game. A destination or hotel VR game can influence customer buying decisions. Cities can use VR game to introduce their own sightseeing spots, as well as local food retailers and different attractions. Experiencing a world through VR might stimulate viewers to travel to its authentic world. VR games provide a full and detailed view of a place. For instance, VR games make tourism destinations more interesting and easier. It offers a detailed view of a place wherever the player wishes to go; therefore, viewers can organise their trip to seeing the real locations of that destination.

Problems with the game

It requires a VR headset, a racked motion controller and a room scale play area, whereas in AR games, a smartphone is usually sufficient and is therefore more accessible and affordable to most players. Using a VR headset can be harmful to the eyes or can cause other health problems. If a player spends an extended period in a VE, they may become addicted to it.

National Geographic Explore VR

The interactive experience has two episodes. The first experience takes the player on an expedition to Antarctica, where they can kayak in icy waters and climb mountains. The player navigates around icebergs in a kayak, climbs a massive ice shelf and survives a raging snowstorm as they search for a lost emperor penguin colony. The player experiences the breathtaking scenery and wildlife, takes pictures and shares them with the world. The second experience is based in the Andes mountains in Peru. The player can learn about the area's history as they trek through the Incan citadel, Machu Picchu, in Peru. The player is immersed in amazing digital reconstructions of the ancient Inca citadel. They also witness mummy worship, raise a cup of sacred chicha and encounter alpacas (Oculus, 2020).

Game type

Simulation VR application of a culture and nature tour. It needs a VR headset, tracked motion controller and room scale play area. It is currently active for playing but it is not free.

Target group

Players who want to make virtual culture and nature tours in Antarctica and Peru.

Marketing benefit

Promotes Antarctica's nature and Machu Picchu's culture. With the help of VR headsets, the player is fully immersed in the virtual culture and nature tour.

Marketing applications

The player discovers the Andes mountains in Peru and navigates around icebergs in a kayak in Antarctica. After seeing the places in a virtual world, the player may want to explore Antarctica and Machu Picchu.

Marketing communication

It is a single player game. Different destinations can be added as new episodes in the game, such as a ballooning tour in Cappadocia or a safari tour in Africa. A virtual culture and nature tour can influence customer buying decisions. VR games provide full and detailed views of a place, a destination or a tour. VR games make touristic tours more interesting and easier to marketing the tours. It offers a detailed view of a place, wherever the player wishes to go.

Problems with the game

Playing the game can be expensive for players because it needs a VR headset, tracked motion controller and room scale play area, and it is not a free application. Also, using a VR headset can be harmful to the eyes or can cause other health problems. If someone spends an extended period in a VE, they may become addicted to it.

Ocean Rift

Ocean Rift is the world's first VR aquatic safari park and it is similar to TheBlu VR application. The player can explore a vivid underwater world full of life, including dolphins, sharks, turtles, orcas, sea snakes, rays, whales, manatees, sea lions and even dinosaurs. Ocean Rift is divided into 12 unique habitats ranging from coral reefs, shipwrecks and lagoons, to the arctic, the deep and prehistoric seas and Atlantis. In the game, the player can swim around, explore and interact with sea creatures across 12 unique habitats. Also, all-star animals in the game have their own artificial intelligence, which means that every visit to Ocean Rift will be a unique experience for the player (Oculus, 2020).

Game type

Aquatic safari park simulation VR application. It needs a VR headset, tracked motion controller and room scale play area. It is currently active for playing but it is not free.

Target group

Players who want to take an aquatic safari in a virtual world.

Marketing benefit

Promotes the seas and oceans and gives information about sea/ocean nature and habitat.

Marketing applications

Players live and dive in a virtual ocean and explore a vivid underwater world full of life, including dolphins, sharks, turtles, orcas, sea snakes, rays, whales, manatees, sea lions and even dinosaurs.

Marketing communication

VR games are much more entertaining and immersive than other marketing communication applications. Playing virtual reality games gives the player the chance to step into a completely different world. This will make the entire tourism experience more realistic than advertisements. It also gives the chance to experience a diving tour in the oceans. If they like the experience in the game, they might buy the real tourism tour.

Problems with the game

For the VR game market to grow, and for VR to prove its value to consumers, prices need to adjust to consumer budgets. A VR headset can be expensive for players. Some VR headsets carry a high price tag, especially for proprietary closed-face designs. Wearing VR headset for long periods produces fatigue and an unsettling feeling of enclosure, as well as dry eyes and strain on eyes. Also, it is not free, so players need to purchase it.

National Archaeological Museum: Live the Past

The player lives the experience of visiting several towns throughout the history of Spain and participate in a VR experience to learn how people lived in different times Spain. The application allows the player to experience historical recreations of the museum's audio visual heritage, the streets and homes of prehistory, protohistory, Roman, Medieval and the modern age. These scenes have been carefully supervised by the teams of curators of the National Archaeological Museum (Oculus, 2020).

Game type

Museum simulation VR application. It needs a VR headset, tracked motion controller and room scale play area. It is currently active for playing and it is free.

Target group

Players who want to explore the history of Spain in a virtual world.

Marketing benefit

Promotes the National Archaeological Museum in Spain. After seeing the museum in a virtual world, players may want to explore the museum or Spain's historical places in the real world.

Marketing applications

The player discovers the history of Spain with the help of VR tools and the player is fully immersed in the virtual museum.

Marketing communication

It is a single player game. Different museums in Spain can be added as new episodes in the game. A virtual museum tour can influence customer buying decisions. Virtual reality games provide full and detailed views and information about a museum, destination or tour. For instance, VR games make museums more interesting, interactive and easier to marketing the museums. It allows players to experience a museum tour who otherwise might not be able to visit (e.g. people with disabilities).

Problems with the game

It requires a VR headset and a tracked motion controller, which might be expensive for the player. Also, using a VR headset can be harmful to the eyes or can cause other health problems. If someone spends an extended period in a VE, they may become addicted to it.

VR games in tourism industry, such as the ones mentioned above, take the users out of the real world and places them in an imaginary one. That is, they may find themselves in a new environment, either real or imaginary, within seconds. What makes it possible now is the use of a head-mounted display or a hand-held controller, which helps them navigate the world they are in. AR does not remove the users from their real environment, but augments their surroundings by adding other elements such as graphics, sounds, videos, text and the like. For example, with VR the users may find themselves in an ocean swimming with sharks, but with AR a shark may pop out of advertisements or tourism brochures. AR offers more freedom to users and more possibilities to marketers because they do not require specialist equipment such as a head-mounted display (Pal, 2017).

The advantages of AR and VR games with examples from the marketing perspective are summarised in Table 12.2. In the game examples, AR games are multiplayer games, but VR games are single player games (see Table 12.2). Most of the games include sightseeing of a city or destination, capture data about customers and increase city or destination awareness. AR games allow communication with other players, interactive marketing and socialisation with players, making a real life city tour fun and exciting. VR games have helped in customer buying decisions by providing a preview of a city or destination as well as giving detailed views and information about the city or destination. A disadvantage of the games are that they can cause addiction to the virtual world. Both game types capture data about the player and increase city or destination awareness. Another disadvantage is that VR games require expensive technological tools and can cause dry eyes and strain on the eyes. VR headsets and tracked motion controllers can also be expensive for everyday users. Mostly, the games' development purpose is for having fun with the game and city and giving information about the destination. An advantage of VR games is that they allow players to experience a museum, destination or city tour, which makes them accessible for tourists with a disability who cannot travel physically but would like to explore different destinations.

Table 12.2 Marketing advantages of AR and VR games: examples from the tourism industry

	Pokémon GO	Tripventure	Explore	Ingress	Rexplorer	Loquiz	Avastusrada	Harry Potter: Wizards Unite	City VR	Vacation Simulator	TheBlu VR	Guided Meditation VR	Nature Treks VR	National Geographic Explore VR	Ocean Rift	National Archaeological Museum
Game type																
AR game	*	*	*	*	*	*	*	*								
VR game									*	*	*	*	*	*	*	*
Player number																
Single player									*	*	*	*	*	*	*	*
Multiplayer					*	*	*									
Both	*	*	*	*				*								
Marketing advantages																
Sightseeing of a city or destination	*	*	*	*	*	*	*	*	*		*	*	*	*	*	*
Help in customer buying decisions									*	*	*	*	*	*	*	*
Gives detailed views of the city or destination		*	*		*				*		*	*	*	*	*	*
Communication with other players	*	*	*	*	*	*	*	*								
Captures data about customers	*	*	*	*	*	*	*	*	*	*	*	*	*	*	*	*
Increases city or destination awareness	*	*	*	*	*	*	*	*	*		*	*	*	*	*	*
Interactive marketing	*	*	*	*	*	*	*	*								
Socialisation with players	*	*	*	*	*	*	*	*								
Preview of a city or destination									*		*	*	*	*	*	*
Making real-life city tours fun and exciting	*	*	*	*	*	*	*	*								
Giving information about historical places		*		*	*	*	*							*		*
Giving information about museums		*	*	*	*	*	*									*
Promotions of tourism businesses	*			*				*								
Problems with the games																
Can cause addiction to the virtual world	*	*	*	*	*	*	*	*	*	*	*	*	*	*	*	*
Requires expensive technological tools									*	*	*	*	*	*	*	*
Can cause dry eyes and strain on eyes									*	*	*	*	*	*	*	*
Can cause accidents	*	*	*	*	*	*	*	*								
Essential technologies for playing the game (during playing)																
Smart phone	*	*	*	*	*	*	*	*								
GPS reception	*	*	*	*	*	*	*	*								
Bluetooth			*													
Internet	*	*	*	*	*	*	*	*								
VR headset									*	*	*	*	*	*	*	*
Tracked motion controller									*	*	*	*	*	*	*	*

(Continued)

Table 12.2 (*continued*)

	Pokémon GO	Tripventure	Explore	Ingress	Rexplorer	Loquiz	Avastusrada	Harry Potter: Wizards Unite	City VR	Vacation Simulator	TheBlu VR	Guided Meditation VR	Nature Treks VR	National Geographic Explore VR	Ocean Rift	National Archaeological Museum
Purpose of the game																
Fun for fun's sake	*	*	*	*	*	*	*	*	*	*	*	*	*	*	*	*
Have fun with city/product	*	*	*	*	*	*	*	*	*		*	*	*	*	*	*
Education		*	*	*	*	*	*		*		*	*	*	*	*	*

Discussion and Conclusion

Different implementations of AR and VR games have emerged owing to the experience gained by the developers, hence there is more creativity regarding applications. It is obvious that players are more engaged and interactive in games compared with display rich media. As for the commercials on TV, viewers have the freedom to fast forward or engage in other activities; yet in game-based marketing, the consumers need to be fully concentrated on the game, hence there is more engagement and avoidance from other distractions such as online chatting. Game-based marketing makes integrated marketing efforts possible; thus, many companies have developed games to increase both the level of interaction and their database. Brand awareness and recall is enhanced through AR and VR games. If a city or destination attracts players, they seek more information about it on their website to make informed decisions. The guidelines to develop AR and VR games in tourism are given in Table 12.3.

Technology is constantly developing, and internet gaming technology is no exception as it enhances its capabilities via the internet. Games lead players to surf the internet, which creates traffic to websites, and therefore opportunities to reinforce brands. In games, users are required to register so that they will be eligible for the prizes given. This helps marketers collect data about their existing or prospective customers. It is also common among gamers to invite friends, which further increases the brand awareness of tourism products. In other words, it functions as 'viral marketing'. Recent studies (Azuma, 2017; Bonetti *et al.*, 2018; Cranmer, 2017; Xu *et al.*, 2017; Xu *et al.*, 2016) in brand impact studies suggest that the fun aspect of the games helps increase brand metrics such as brand awareness, message association and purchase intent. That is to say, it increases the likelihood of remembering the brand and the product itself after playing a game, and also associates specific brand(s) attributed within the game.

In this chapter, 16 AR and VR games (Pokémon GO, Tripventure, Explore, Ingress, REXplorer, Loquiz, Avastusrada, Harry Potter: Wizards Unite, City Explorer, Vacation Simulator, TheBlu VR, Guided Meditation VR, Nature Treks VR, National Geographic Explore VR, Ocean Rift, National Archaeological Museum) applications were analysed (see Table 12.2). There are few examples of AR and VR games in the tourism industry. Because AR and VR games in the tourism industry are at the development stage and these game applications are new for the tourism industry, AR and VR applications are mostly used in tourism guides and advertisements.

The properties of the games can be summarised as follows: AR games are multiplayer games, but VR games are single player. AR games are mostly location-based applications. Most of the games have sightseeing of a city or destination, capture data about customers and increase city or destination awareness. This shows the games have marketing potential in the tourism industry. AR games have communication with other players, interactive marketing and socialisation with players, making real life city tours fun and exciting. This means AR games have advantages in interactive communication with customers and entertaining the tourist. VR games help in customer buying decisions and preview of the city or destination as well as giving detailed views of city or destination. According to this result, VR games can help with the tourist buying decision stage. The AR games allow the player to play in the real world so that the player has the opportunity of finding entertaining information during a tourist destination visit or museum trip. Therefore, AR games are useful for making museum and destination trips fun. AR games are played in a real tourism venue. VR games are played in a room environment with the necessary technological tools. They also provide the opportunity to experience a museum, destination or city tour for tourists with disabilities who cannot travel physically but would like to explore different destinations.

The characteristics of an efficient game, for companies specialised in game design and development stages, include the following, which might be taken as a basis in the creation and implementation of the games (Gurau, 2007; Zodal, 2017):

- Carefully selected game genre to attract the targeted audience.
- Relating the core brand message to the game.
- Integration of the brand message into the game, which makes the game more effective and the brand more appealing.
- Carefully levelled game phases, from the simplest to the most complex one, to keep the players engaged by gradually increasing the challenge and make the internalisation of the brand higher.
- Requiring registration using their personal data to post and enter in competitions.
- Inclusion of communication tools to promote experience sharing with other people in their lives, that is, the viral marketing aspect.

Table 12.3 Guidelines for developing AR and VR games in tourism

Area	Guideline for games	Description
Players	Targeting player types: In the tourism industry, game developers should consider the four player types and target them. The model that Bartle (1996) obtained as a result of his work depends on four player categories (achievers, explores, killers and socialisers). Although player types are divided into four, it is difficult to make certain restrictions between player types. Each player can be a little bit of each (Bartle, 1996).	Achievers: This player tries to achieve targets in the game. The targets of the game can be collecting points, passing levels and collecting badges. Achievers want to show their skills and strive for 100% completion rates. Explorers: This player is focused on interactions and wants to explore new methods and interesting or new ways. Explorers want to see new things and discover new secrets. They do not care about points and prizes. For them, discovery is the prize. These players prefer mostly adventure games. Killers: Killers are focused on their rivals; their only aim is to beat their rivals. Their main aims are to win the game and to make sure other players lose. Socialisers: Socialisers are interaction-focused players; the game is a tool for them to socialise. They do not target like defeating, gathering points or discovering. Having a nice time with other people is important for them. Simple games and low competition games are preferred by these players (Bartle, 1996).
Stages at which the game aims to market or promote the tourism industry	Before travelling, during travelling and after travelling games can give information or details about tourism products at different stages.	Before travelling VR and AR games can be used to give information and details about cities, destinations or tourism services as well as help with the buying decision of the tourist at this stage. During travelling: VR and AR games can be used to provide their customers with experiences that are remarkable and beautiful while motivating the consumption levels of tourists to further increase the use of the product or service. City- or destination-based treasure hunt games can be used at this stage. After travelling: With VR and AR games businesses can increase their customer loyalty, satisfaction and can develop a customer database.
Tourism marketing applications in games	Games might be used as a marketing communication tool for the tourism industry.	Game developers focused on using these marketing applications in games: sharing on social media, sharing with friends, the opportunity of making in-game purchases, providing promotion opportunities, giving information about the city, destination or businesses in the game, interactive communication with other players, giving rewards and discounts from tourism businesses.

Table 12.3 Guidelines for developing AR and VR games in tourism

Area	Guideline for games	Description
Gamification theories	AR and VR games can be developed based on gamification theories. Integrating the games according to the four game theories can increase the game's advantages in tourism. Game theories such as self-determination theory, hook theory, Fogg behaviour theory, and flow theory should be considered by game developers.	Self-determination theory: Having a sense of choice by choosing behavioural preferences that meet the innate psychological needs of individuals and enabling them to create their own intrinsic motivation is called self-determination theory. This theory has three realms: autonomy, competence and relatedness (Deci *et al.*, 1989) Hook theory: Hook theory is a four-staged theory that has been developed by Nir Eyal, with the aim of ensuring maintenance of behaviours by directing people to specific behaviours. The four stages that take place in hook theory are trigger, action, variable rewards and investment (Eyal & Hoover, 2014). Fogg behaviour theory: This theory was developed by Professor BJ Fogg in 2009. This theory consists of three elements: motivation, ability and triggers (Fogg, 2009). Flow theory: The main aim of flow theory in games is to create an opportunity to have fun for players so they enjoy games, providing an increase of inner motivations (Csikszentmihalyi, 1990).

- Offering competitions with prizes to encourage players to register and return to the game.
- Providing visible scoreboards that encourage players to perform better, be more competitive and invite more players to the competition.

In the tourism industry, to strengthen the development of marketing communication and branding in games, the main goals to be achieved are the following:

- To provide a fun experience with content related to the city, destination, museum, culture or nature tour.
- To lead consumers to purchase or complete trials.
- To increase the product brand awareness on- and offline by creating a need and public interest for a future visit.
- Utilising viral games to attract the target audience and supporting such campaigns with the help of other marketing and advertising tools.

For governments and local authorities who wish to turn their cities/regions into model branded and popular destinations, the following suggestions are outlined:

- The potential tourist and game players must be identified.
- A themed game, underlining touristic information about the destinations/cities must be created.

Further research on the following areas is recommended:

- Integration of VR and AR game technologies into tourism marketing, including VR and AR advergames for tourism destinations and businesses.
- Boosting the tourist experience by analysing the demands of tourists about VR and AR game applications and devices.
- Improving marketing applications in games.

Finally, we should be aware of a dynamic environment that evolves rapidly as new applications enter the marketplace, hence increasing the number of AR and VR games available. Therefore, ongoing follow-up surveys might help enlarge the sample size and gain more insights into the trends for further analysis. Nevertheless, this study sheds light on VR and AR game application technology by providing a crucial overview, which may help revolutionise the way people travel.

References

148apps.com Available at: http://www.148apps.com/app/442671519/ (accessed 6 April 2018).

Appdedice.com (2018) Tripventure. Available at: https://www.appdecide.com/app/ios/tripventure/442671519/ (accessed 10 April 2018).

Augray (2018) Augmented reality vs virtual reality in gaming. Available at: https://augray.com/blog/augmented-reality-virtual-reality-gaming/ (accessed 20 October 2018).

Azuma, R.T. (1997) A survey of augmented reality. *Teleoperators and Virtual Environments* 6 (4), 355–385.
Azuma, R. (2015) Location-based mixed and augmented reality storytelling. W. Barfield (ed.) *Fundamentals of Wearable Computers and Augmented Reality* (pp. 259–276) (2nd edn). Boca Raton, Florida: CRC Press.
Azuma, R.T. (2016) The most important challenge facing augmented reality. *Teleoperators and Virtual Environments* 25 (3), 234–238.
Azuma, R.T. (2017) Making augmented reality a reality. Proceedings of OSA Imaging and Applied Optics Congress, 25–29 June 2017, San Francisco, CA.
Azuma, R., Baillot, Y., Behringer, R., Feiner, S., Julier, S. and MacIntyre, B. (2001) Recent advances in augmented reality. *IEEE Computer Graphics and Applications* 21 (6), 34–47.
Bartle, R. (1996) Hearts, clubs, diamonds, spades: Players who suit MUDs. *Journal of MUD Research* 1 (1). Available at: http://www.mud.co.uk/richard/hcds.htm (accessed 17 December 2020).
Berryman, D.R. (2012) Augmented reality: A review. *Medical Reference Services Quarterly* 31 (2), 212–218.
Bonetti, F., Pantano, E., Warnaby, G., Quinn, L. and Perry, P. (2018) Augmented reality in real stores: Empirical evidence from consumers' interaction with AR in a retail format. 4th International AR & VR Conference, 21–22 June, Manchester.
Buhalis, D., Harwood, T., Bogicevic, V., Viglia, G., Beldona, S. and Hofacker, C. (2019) Technological disruptions in services: Lessons from tourism and hospitality. *Journal of Service Management* 30 (4), 484–506.
Carmigniani, J., Furht, B., Anisetti, M., Ceravolo, P., Damiani, E. and Ivkovic, M. (2011) Augmented reality technologies, systems and applications. *Multimedia Tools and Applications* 51 (1), 341–477.
Castro, J.C., Quisimalin, M., Cordova, V.H., Quevedo, W.X., Gallardo, C., Santana, J. *et al.* (2017) Virtual reality on e-tourism. International Conference on IT Convergence and Security (ICITCS). 25–28 September, Singapore: Springer, 86–97.
Chen, H. (2008) Mobile games in tourism. Available at: http://64.233.183.104/ search?q= cache:cbiSgvWR4oJ:interactive.usc.edu/members/hchen/2008/10/gaming_tourism_lessons_from_ev.html+%22mobile+games+in+tourism%22&hl=en&ct=clnk&cd=5 (accessed 6 April 2018).
Çeltek, E. (2015) Augmented reality advertisements in tourism marketing. In N. Takıran and R. Yılmaz (eds) *Handbook of Research on Effective Advertising Strategies in the Social Media Age* (pp. 125–146). Hershey, PA: IGI Global.
Cheong, R. (1995) The virtual threat to travel and tourism. *Tourism Management* 16 (6), 417–422.
Chou, T-L. and ChanLin, L-J. (2012) Augmented reality smartphone environment orientation application: A case study of the Fu-Jen University mobile campus touring system. *Procedia – Social and Behavioral Sciences* 46 (2012), 410–416.
Chung, N., Han, H. and Joun, Y. (2015) Tourists' intention to visit destination: Role of augmented reality applications for heritage site. *Computers in Human Behavior* 50 (2015), 588–599.
Craig, A.B. (2013) *Understanding Augmented Reality, Concepts and Applications* (1st edn). San Francisco: Morgan Kaufmann.
Cranmer, E.E. (2017) Developing an augmented reality business model for cultural heritage tourism: The case of Geevor Museum. Unpublished doctoral thesis, Department of Operations, Technology, Events and Hospitality Management, Manchester Metropolitan University.
Csikszentmihalyi, M. (1990) *Flow: The Psychology of Optimal Experience.* New York: Harper Perennial.
Deci, E.L., Connell, J.P. and Ryan, R.M. (1989) Self-determination in a work organization. *Journal of Applied Psychology* 74 (4), 580–590.

Eyal, N. and Hoover, R. (2014) *Hooked: How to Build Build Habit Forming Products.* Wood Dale, IL: Penguin Group.

Eyles, M. and Eglin, R. (2008) Ambient games, revealing a route to a world where work is play? *International Journal of Computer Games Technology.* Available at: https://dl.acm.org/doi/pdf/10.1155/2008/176056 (accessed 17 December 2020).

Fino, E., Martín-Gutiérrez, J., Fernández, M. and Davara, E. (2013) Interactive tourist guide: Connecting Web 2.0, augmented reality and QR codes. *Procedia Computer Science* 25 (2013), 338–344.

Fogg, B. (2009) A behavior model for persuasive design. Proceedings of the 4th International Conference of Persuasive Technology, 26–29 April. Claremont, CA.

Fong, A. (2011) Augmented reality apps turn smartphones into digital tour guides. Available at: http://business.financialpost.com/2012/07/11 /augmented-reality-apps-turn-smart-phones-into-digital-tour-guides/?_lsa=c75a-1118 (accessed 20 September 2018).

Fritz, F., Superregui, A. and Linaza, M. (2005) Enhancing cultural tourism experiences with augmented reality technologies. The 6th International Symposium on Virtual Reality, Archaeology and Cultural Heritage, VAST, November, Pisa, Italy, 1–5.

Gibson, A. and O'Rawe, M. (2018) Virtual reality as a travel promotional tool: Insights from a consumer travel fair. In T. Jung and M. TomDieck (eds) *Augmented Reality and Virtual Reality* (pp. 93–107). Progress in IS. Cham: Springer.

Google Play (2018) City Explorer. Available at: https://play.google.com/store/apps/details?id=com.PeterLabickDevelopment.CardboardCityExplorer&hl=fil (accessed 10 April 2018).

Google Play (2020) Harry Potter: Wizards Unite. Available at: https://play.google.com/store/apps/details?id=com.nianticlabs.hpwu.prod&hl=tr (accessed 10 April 2020).

Gura, S., Panajoti, V.H. and Gura, K.S. (2018) The relationship between advergame and brand loyalty. *International Journal of Economics, Commerce and Management* VI (11), 183–206.

Gurau, C. (2007) *Virtual Reality Applications in Tourism, Information and Communication Technologies in Support of the Tourism Industry* (pp. 180–197). Hershey PA: IGI Global.

Guttentag, D.A. (2010) Virtual reality: Applications and implications for tourism. *Tourism Management* 31 (2010), 637–651.

Han, D-I., Jung, T. and Gibson, A. (2014) Dublin AR: Implementing augmented reality in tourism. In Z. Xiang and I. Tussyadiah (eds) *Information and Communication Technologies in Tourism* (pp. 511–523). New York: Springer International Publishing.

Hassan, A. and Rahimi, R. (2016) Consuming 'innovation' in tourism: Augmented reality as an innovation tool in digital tourism marketing. In N. Pappas and I. Bregoli (eds) *Global Dynamics in Travel, Tourism and Hospitality* (pp.13–147). USA: IGI Global.

He, Z., Wu, L. and Li, X. (2018) When art meets tech: The role of augmented reality in enhancing museum experiences and purchase intentions. *Tourism Management* 68 (2018), 127–139.

Höllerer, T.H. and Feiner, S.K. (2004) Mobile augmented reality. In H. Karimi and A. Hammad (eds) *Telegeoinformatics: Location-Based Computing and Services* (pp. 221–262). Boca Raton, Florida: CRC Press.

https://cityvr.com/ (2020) City VR, Available at: https://cityvr.com/images/sanfrancisco/Screenshot_6.jpg (accessed 10 November 2020).

https://www.ionicsmusic.com/trip-venture (2020) Trip Venture app, Available at: https://static.wixstatic.com/media/6e61c1_471636186cce4357ac78c3f8d185dc40.jpg/v1/fill/w_968,h_481,al_c,q_85/6e61c1_471636186cce4357ac78c3f8d185dc40.webp (accessed 10 November 2020).

https://www.youtube.com (2020) Augmented Reality Postcards - Dubai (AR Tourism), Available at: https://www.youtube.com/watch?v=650jnMilsf8 (accessed 10 November 2020).

https://www.youtube.com (2020) DiscoverHongKong-AR by XS2TheWorld Available at: https://www.youtube.com/watch?v=Iw15uXQ4puY, (accessed 10 November 2020).

https://www.youtube.com (2020) National Geographic Explore | Oculus Quest, Available at: https://www.youtube.com/watch?v=D1VILv6Db7I (accessed 10 November 2020).

https://www.youtube.com (2020) Starbucks Cup Magic for Valentine's Day, Available at: https://www.youtube.com/watch?v=8nvqOzjq10w (accessed 10 November 2020).

https://www.youtube.com (2020) Trailer theBlu. Available at: https://www.youtube.com/watch?v=A8XIYjNoovo (accessed 10 November 2020).

Huang, Y.C., Backman, K.F., Backman, S.J. and Chang, L.L. (2016) Exploring the implications of virtual reality technology in tourism marketing: An integrated research framework. *International Journal of Tourism Research* 18, 116–128.

Ivanova, I. (2016) Augmented reality games for increasing physical activity. Unpublished master's thesis, Tallinn University.

Kennedy-Eden, H. and Gretzel, U. (2012) A taxonomy of mobile applications in tourism. *E-Review of Tourism Research* 10 (20), 47–50.

Koikawa, K. (2016) Ideas to boost tourism: From William Gibson's Spook Country to 'Pokémon Go,' and mixed reality. *Journal of Global Tourism Research* 1 (2), 105–109.

Kounavis, C., Kasimati, A. and Zamani, E. (2012) Enhancing the tourist experience through mobil augmented reality: challenges and prospects. *International Journal of Engineering Business Management* 4 (10), 1–6.

Kourouthanassis, P., Boletsis, C., Bardaki, C. and Chasanidou, D. (2015) Tourists responses to mobile augmented reality travel guides: The role of emotions on adoption behavior. *Pervasive and Mobile Computing* 18 (2015), 71–87.

Kuniavsky, M. (2006) REXplorer: Magic as historical explanation metaphor. Available at: http://www.orangecone.com/archives/2006/11/rexplorer_magic_1.html (accessed 10 October 2018).

Lashkari, A., Parhizkar, B. and Mohamedali, M. (2010) Augmented reality tourist catalogue using mobile technology. Second International Conference on Computer Research and Development, 7–10 May 2010, Kuala Lumpur, Malaysia.. Washington, DC: IEEE, 121–125.

LaValle, S.M. (2019) *Virtual Reality*. Cambridge: Cambridge University Press.

Lee, K. (2012) Augmented reality in education and training. *Tech Trends* 56 (2), 13–21.

Lifehacker.com (2018) How Ingress, Google's real-world smartphone game, got me out of my shell. Available at: https://lifehacker.com/how-ingress-googles-real-world-smartphone-game-got-me-1710320867 (accessed 15 May 2018).

Linaza, M., Marimon, D., Carrasco, P., Alvarez, R., Montesa, J., Aguilar, S.R. and Diez, G. (2012) Evaluation of mobile augmented reality applications for tourism destinations. In M. Fuchs, F. Ricci and L. Cantoni (eds) *Information and Communication Technologies in Tourism* (pp. 260–271). Wien: Springer-Verlag.

Lopatina, I. (2005) Context-aware mobile gaming. Unpublished master's thesis, University of Troms.

Marimon, D., Sarasua, C., Carrasco, P., Alvarez, R., Montesa, J., Adamek, T., Romero, I., Ortega, M. and Gasco, P. (2018) Mobi AR: Tourist experiences through mobile augmented reality. Available at: http://www.researchgate.net/publication/228979424_MobiAR_Tourist_Experiences_through_Mobile_Augmented_Reality (accessed 10 October 2018).

McCall, R., Wetzel, R., Löschner, J. and Braun, A-K.K. (2011) Using presence to evaluate an augmented reality location aware game. *Personal and Ubiquitous Computing* 15 (1), 25–35.

Moorhouse N., TomDieck M.C. and Jung T. (2018) Technological innovations transforming the consumer retail experience: A review of literature. In T. Jung and M. TomDieck (eds) *Augmented Reality and Virtual Reality* (pp. 133–143). Progress in IS. Cham: Springer.

Niantic and Nintendo (2016) Pokémon Go 0.31.0. Paolo Alto, CA, USA.
Niantic (2020) Harry Potter: Wizards Unite. Available at: https://play.google.com/store/apps/details?id=com.nianticlabs.hpwu.prod&hl=tr (accessed 10 April 2020).
Oculus (2020a) The BluVR. Available at: https://www.oculus.com/experiences/gear-vr/2221544801220466/?ranking_trace=0_2221544801220466_SKYLINEWEB_1tBMACvaHSAPviiB3 (accessed 10 April 2020).
Oculus (2020b) Nature Treks VR. Available at: https://www.oculus.com/experiences/quest/2616537008386430/?ranking_trace=0_2616537008386430_SKYLINEWEBQUESTSEARCH_1uewBPQFikwHxp2tp (accessed 10 April 2020).
Oculus (2020c) National Geographic Explore VR. Available at: https://www.oculus.com/experiences/quest/2046607608728563/ (accessed 10 April 2020).
Oculus (2020d) Ocean Rift. Available at: https://www.oculus.com/experiences/quest/2134272053250863/ (accessed 10 April 2020).
Oculus (2020e) National Archaeological Museum: Live the Past. Available at: https://www.oculus.com/experiences/go/1144207145676653/?ranking_trace=0_1144207145676653_SKYLINEWEB_1Du3iI3sbQFr09rXi (accessed 10 April 2020).
Pal, G. (2017) Artificial intelligence and augmented reality (AR) vs virtual reality (VR) – Which is best for future. Available at: https://howtechhack.com/artificial-intelligence-augmented-reality-ar-vs-virtual-reality-vr-best-future/ (accessed 20 October 2018).
Road to VR (2018) Hands-on: 'Vacation Simulator' is a more expansive sequel that dives deeper. Available at: https://www.roadtovr.com/hands-on-vacation-simulator/ (accessed 10 April 2018).
Rouse, M. (2016a) Virtual reality gaming (VR gaming). Available at: https://whatis.techtarget.com /definition/virtual-reality-gaming-VR-gaming (accessed 20 April 2018).
Rouse, M. (2016b) Augmented reality gaming (AR gaming). Available at: https://whatis.techtarget.com/ definition/augmented-reality-gaming-AR-gaming (accessed 20 April 2018).
Russell, M. (2012) 11 Amazing augmented reality ads. http://www.businessinsider.com/11-amazing-augmented-reality-ads-2012-1?op= 1 (accessed 10 March 2017).
Schrier, K. (2006) Using augmented reality games to teach 21st century skills. *ACM SIGGRAPH 2006 Educators Program* (SIGGRAPH 2006). New York: ACM, article 15. DOI: https://doi.org/10.1145/1179295.1179311.
Sekhavat, Y.A. (2016) KioskAR: An augmented reality game as a new business model to present artworks. *International Journal of Computer Games Technology*. DOI: 10.1155/2016/7690754.
Selvam, A., Yap, T.T-V., Ng, H., Tong, H-L. and Ho, C-C. (2016) Augmented reality for information retrieval aimed at museum exhibitions using smartphones. *Journal of Engineering and Applied Sciences* 100 (3), 635–639.
Sing, P. and Pandey, M. (2014) Augmented reality advertising: An impactful platform for new age consumer engagement. *IOSR Journal of Business and Management (IOSR-JBM)* 16 (2), 24–28.
Stella Artois Le Bar Guide augmented reality. Available at: https://www.youtube .com/watch?v=sTERI1s-UyA (accessed 20 September 2018).
Tavakoli, R. and Mura, P. (2015) Journeys in second life – Iranian Muslim women's behaviour in virtual tourist destinations. *Tourism Management* 46 (2015), 398–407.
Taylor, B. (2013) Augmented reality applications – From the consumer to the business. Available at: http://www.creativeguerrillamarketing.com/augmented-reality/augmentedreality-applications-from the -consumer-to-thebusiness/ (accessed 6 April 2018).

Tokareva, J. (2018) The difference between virtual reality, augmented reality and mixed reality. Available at: https://www.forbes.com/sites/quora/2018/02/02/the-difference-between-virtual-reality-augmented-reality-and-mixed-reality/#2738f182d07c (accessed 20 September 2018).

TomDieck, M.C. and Jung, T. (2017) Value of augmented reality at cultural heritage sites: A stakeholder approach. *Journal of Destination Marketing and Management* 6 (2017), 110–117.

Tromp, P. (2017) How virtual reality will revolutionize the hospitality industry. Available at: https://www.hospitalitynet.org/opinion/4080737.html (accessed 10 October 2018).

Tussyadiah, I.P., Wang, D., Jung, T.H. and TomDieck, M.C. (2018) Virtual reality, presence, and attitude change: Empirical evidence from tourism. *Tourism Management* 66 (2018), 140–154.

Wagner-Greene, V., Wotring, A.J., Castor, T., Kruger, J., Mortemore, S. and Dake, J.A. (2017) Pokémon GO: Healthy or harmful? *American Journal of Public Health* 107 (1), 35–36. DOI: 10.2105/AJPH.2016.303548.

Wang, X., Kim, M., Love, P. and Kang, S-C. (2013) Augmented reality in built environment: Classification and implications for future research. *Automation in Construction* 32 (2013), 1–13.

Wasko, C. (2013) What teachers need to know about augmented reality enhanced learning. *Tech Trends* 57 (1), 17–21.

Weber, J. (2014) Information and communication technologies in tourism 2014 ENTER, Dublin, Ireland, R. Baggio, M. Sigala, A. Inversini and J. Pseonen (eds) January 21–24. London: Springer.

Weber, J. (2016) Designing engaging experiences with location-based augmented reality games for urban tourism environments. Unpublished doctoral thesis, Bournemouth University.

Williams, A.P. and Hobson, J.S.P. (1995) Virtual reality and tourism: Fact or fantasy? *Tourism Management* 16 (6), 423–427.

Xu, F., Buhalis, D. and Weber, J. (2017) Serious games and the gamification of tourism. *Tourism Management* 60 (2017), 244–256.

Xu, F., Tian, F., Buhalis, D., Weber, J. and Zhang, H. (2016) Tourists as mobile gamers: Gamification for tourism marketing. *Journal of Travel & Tourism Marketing* 33 (8), 1124–1142.

Yovcheva, Z., Buhalis, D. and Gatzidis, C. (2013) Engineering augmented tourism experiences. In L. Cantoni and Z. Xiang (eds) *Information and Communication Technologies in Tourism* (pp. 24–35). Berlin Heidelberg: Springer.

Yung, R. and Khoo Lattimore, C. (2017) New realities: A systematic literature review on virtual reality and augmented reality in tourism research. *Current Issues in Tourism* 22 (17), 2056–2081. DOI: 10.1080/13683500.2017.1417359.

Zodal (2017) The lure of the advergame. Available at: http://www.zodal.com/pdf/Advergames-Zodal.pdf (accessed 1 July 2018).

13 Conclusion

Feifei Xu and Dimitrios Buhalis

Gamification has become the focus of attention in an increasing number of fields, including business, education and health care (Xu *et al.*, 2017). Serious gaming is widely used in a range of practices, primarily for training purposes. The ability of gamification to provide content, in an entertaining and engaging way, combining with the ability to scale the application to many gamers adds significant value to various industries. Gradually, the ability of smart systems to connect various networks and players will support team gaming, competitions and collaboration efforts, based on gaming platforms (Buhalis, 2020; Buhalis *et al.*, 2019). Networks and clusters of activities, such as wine tourism and routes, for example, can use gamification to bring all stakeholders together to perform collaborative activities (Brás *et al.*, 2013). This will encourage individuals to compete or collaborate with others in solving problems, designing solutions or addressing societal problems, perhaps in real time (Buhalis & Sinarta, 2019). Artificial intelligence and machine learning will also analyse patterns of behaviour, preferences and motives. This can be correlated with contextual information providing proactive and reactive solutions using gaming platforms (Buhalis & Foerste, 2015). Ultimately, gaming can support eDestination Marketing in the future (Mistilis *et al.*, 2014).

Academic literature, however, has yet to catch up with industry practices, although it is worth noting that there is emerging academic research in this field. Some best practices are worth looking into and certainly provide us with thoughtful insights into how gamification can be applied to the tourism, hospitality and other fields. The chapters in this book vary from theoretical foundations on gaming, gamification to more practical examples of gamification applications used for marketing in destinations, restaurants and airlines. However, many questions still need to be answered.

How to Design a Meaningful Experience for Tourism Gamification

Aparicio *et al.* (2012) developed a framework of gamification, using intrinsic motivational forces. They suggest four steps: (1) identify the reasons for using gamification; (2) find out what intrinsic motivation can be addressed; (3) identify useful game elements; and (4) consider evaluating the framework in the applied system. However, as Seaborn and Fels (2015) point out, the framework has not been applied yet, therefore, more research work is needed. Burke (2014) proposes a 'player experience design process', which consists of seven steps, namely: (1) business outcomes and success metrics; (2) potential users; (3) player goals; (4) engagement model; (5) play space and journey; (6) test; and (7) repeat. Werbach and Hunter (2012) suggest a design framework with that includes 'define-delineate-describe-devise-deploy'. Yet, the above mentioned frameworks have been criticised by Bulencea and Egger (2015: 58) as they argue that gamification design should not be a linear process. They suggest that 'It should rather be a multi-stage design approach that aims to enhance experiences as the nature of experience design process is non-linear and unclear'. This view is supported by Tung and Ritchie (2011), highlighting that the formation of experiences is a complicated process. In recognition of the complicated formation process of experiences, Huotari and Hamiri (2012) argue that designed experiences cannot be assured for everyone.

The MAPS-AIM model discussed in Chapter 3 (Williams) identifies the 11 components of gamified experiences (namely: mechanic, aesthetic, poetic, technologic, dynamic, and hedonic-eudaimonic, schematic, referent, audience, intention and metric), which identified the core process of game design. Williams believes that the design process contains three related circles, which include the content circle, experience circle and connection circle. Nevertheless, more attention should be given to exploring the design framework of meaningful gamification, particularly the application and validation of those frameworks, taking into consideration the nature of tourist experiences. Chapter 2 (Parapnos & Michopoulou) also explores how game mechanisms can be applied by businesses. They identify popular game mechanisms that can be used in gamification. Research on leisure experiences usually includes the following aspects: fantasy and fun (Holbrook & Hirschman, 1982); escape and relaxation (Beard & Ragheb, 1983); entertainment (Farber & Hall, 2007; Pine & Gilmore, 1999); and novelty and surprise (Dunman & Mattila, 2005). Tourism as an experience economy offers a great test bed for developing gamification applications. Tourist experiences are full of individual moments of significance and can be emotional and subjective (Kim *et al.*, 2012; Uriely, 2005). Therefore, the design of frameworks should consider and adapt to different dimensions of tourism. Designing meaningful gamified systems is challenging and needs game designers to decide which activities in tourism and hospitality are suitable for being 'gamified' (Nicholson, 2012; Robertson, 2010). Thus, the prerequisite of implementing a gamified system is a transparent communication of the purpose to support the voluntary participation of players. Perhaps, future work could examine different design frameworks of a fully-fledged game and a gamified system, and compare how this might influence tourist experiences at different stages.

Understanding Tourist Players

Xu *et al.* (2015) identify the different motivations of players of tourism-related games, including curiosity, exploring the destination, socialising, a fun and fantasy experience, challenge and achievement (see Figure 13.1). Game player motivation can be divided into intrinsic and extrinsic motivation. Xu *et al.* (2016: 9) defined the intrinsic motivation of game play as 'playing games for the pure enjoyment of game play and for the hedonistic value of game play'. Intrinsic motivation is argued to have a more significant effect than extrinsic motivation, as it exists within individuals and links the relations between them and their activities more closely (Ryan & Deci, 2000). As shown in the example in Chapter 5 (Shen & Joppe), Ontario's domestic campaign, 'Where Am I', tried to satisfy the different levels of players' intrinsic motivation. This riddle game triggered players' curiosity. People can explore the destination information on the website, read comments and have fun experiences. Players that correctly answer the riddle have a sense of personal achievement. The detailed analysis of Spanish examples in Chapter 9 (Garrido-Pintado) shows how mobile gaming enables users to play games when they are travelling. More and more games are using augmented reality techniques, or combining location-based, augmented reality or virtual reality techniques (Yovcheva *et al.*, 2014). These games are specifically targeting younger players by providing the latest information and enriching visiting experiences for tourist players.

However, even within the tourist market, players may have different requirements, as their motivation may be different. Chapter 10 (Williams) demonstrates that the motivation of tourist players differ in the cities/countries visited, the time played and the knowledge about the tourist attraction gained, and so on. He further identifies three types of tourist

Figure 13.1 Motivation of tourist players
Source: Xu *et al.* (2016)

players, namely: serious leisure players, transitional leisure players and casual leisure players. Understanding how users engage with technology will provide a clear understanding of their requirements (Fan *et al.*, 2019). Researchers also suggest gender, age, experience and personality influence the motivation for playing (Xu *et al.*, 2016). Therefore, future work should look at the different types of tourist players and the influence of social demographic factors on their motivation. The six-fold tourist typology portraying tourist behavioural patterns in using technology to connect while experiencing the destination provides a solid basis for the adoption of gamification (Fan *et al.*, 2019). This area certainly needs further exploration.

How Can Gamification Benefit Tourism

Gamification in tourism has two functions primarily. The outward function mainly relates to tourist experiences and tourism marketing, while the inward function is associated with employee onboarding, management and training. Currently, more research is focused on the outward function of gamification, with particular attention on tourism marketing (Shen *et al.*, 2020; Xu *et al.*, 2016). Less research has been carried on the inward function of gamification. The outward function of gamification can benefit tourism in all three stages (before, during and after) of travel experiences. Xu *et al.* (2017) summarised the benefits of tourism gamification as outline in Table 13.1.

The case studies in Chapters 5, 6, 7 and 9 demonstrate the application of gamification in different sectors in tourism industry, including destinations, airlines and restaurants. Examples in Austria and Spain demonstrate that gamification can be used as a marketing tool, contributing to brand awareness and brand recognition. These chapters provide the base for more empirical research, either quantitative or qualitative, to examine how effective gamification can influence tourist experiences, destination marketing and employee training.

How AI Gaming Could Influence Gamification

AI (artificial intelligence) is influencing every aspect of our lives, including the gaming industry. In videogames, AI is used to generate intelligent behaviours, primarily in non-player characters, often simulating human-like intelligence. For example, some of the well-known chess games and board games can beat the best human players. In action games, AI opponents can listen for player movements, look for footprints, act with random behaviour with human-like emotions or even take cover when a human opponent fires.

The commonly used techniques, such as decision tree and path finding, draw upon existing methods from AI. Some researchers argue that AI

Table 13.1 Benefits of tourism gamification

	When	Benefits	How
Outward function of gamification	Before	• Increase brand awareness, inspiration/dream. • Generate interests. • Information conversion to sales/purchase.	Tourism-specific games (online games; mobile games) feed tourism information to potential tourists, generate interest, stimulate visiting inspiration, increase brand awareness; virtual reality/3D technology as a technological tool supports these functions.
	During	• Engagement/interaction. • Enhance experiences. • Entertainment tool. • Communication tool with locals or fellow travellers.	Location-based games encourage on-site engagement with the destination, augmented reality games interact the player with the real surrounding tourism attractions; gaming as an entertainment tool to kill time and as a communication tool with locals and fellow travellers.
	After	• Loyalty/repeat visit. • Share experiences. • Recall memories. • Advocate/ambassador.	Online games recall memories, encourage sharing experiences, inviting friends, advocate the destination, rewards/coupons in the game and gamified loyalty programmes encourage repeat visit.
Inward function of gamification		• Employee management. • Employee training. • Generate fun interests.	Use of games to familiarise the employee, use for staff training.

Source: Xu *et al.* (2017)

gaming may often not constitute 'true AI', as such techniques do not necessarily facilitate computer learning or fulfil other standard criteria; many games only use simple predetermined patterns. Some games can use AI to examine player behaviour and change their responses to create more challenging opponents. They provide ground for simulations, constructed environments and tests of realism that are the foundation of virtual experiences. Gamified AI algorithms can examine players' behaviour, patterns and preferences, and respond to individual motivations (intrinsic or extrinsic motivation) in the gamified system. Based on this, it can also make personalised recommendations in the system, making it more personal and fitted to the individual's needs. This is an area that is definitely worth exploring, particularly with the emerging trend of AI in a wider context.

In terms of tourism application, in the future, games will be used as a platform that can collect personal information and behaviour patterns and preferences through AI, which can be connected with other smart devices and predict proactive and reactive solutions for e-destination marketing and management (Buhalis & Foerste, 2015; Mistilis *et al.*, 2014).

Conclusions

The concept of tourism gamification is a relatively recent arrival in the context of tourism, but is becoming ever more popular. The theories and cases discussed in this book provide us with some insightful ideas on how tourism can benefit from gamification. From current studies, we can see that research is mainly centred on the outward function (customer engagement) rather than the inward (employee training) implications of gamification. Therefore, future work could also focus on the inward function of gamification. It is critical to appreciate the range of opportunities gamification is bringing into the tourism industry and work closely with different case studies to develop a comprehensive approach.

References

Aparicio, A.F., Vela, F.L.G., Sánchez, J.L.G. and Montes, J.L.I. (2012) Analysis and application of gamification. Proceedings of the 13th International Conference on Interacción Persona-Ordenador. INTERACCION'12, ACM, Elche, 3–5 October, Spain, p. 17.

Beard, J. and Ragheb, M.G. (1983) Measuring leisure motivation. *Journal of Leisure Research* 15 (3), 219–228.

Buhalis, D. (2020) Technology in tourism-from information communication technologies to eTourism and smart tourism towards ambient intelligence tourism: A perspective article. *Tourism Review* 75 (1), 267–272

Buhalis, D., Harwood, T., Bogicevic, V., Viglia, G., Beldona, S. and Hofacker, C. (2019) Technological disruptions in services: Lessons from tourism and hospitality. *Journal of Service Management* 30 (4), 484–506.

Buhalis, D. and Foerste, M. (2015) SoCoMo marketing for travel and tourism: Empowering co-creation of value. *Journal of Destination Marketing & Management* 4 (3), 151–161.

Buhalis, D. and Sinarta, Y. (2019) Real-time co-creation and nowness service: Lessons from tourism and hospitality. *Journal of Travel & Tourism Marketing* 36 (5), 563–582. DOI: 10.1080/10548408.2019.1592059.

Bulencea, P. and Egger, R. (2015) *Gamification in Tourism: Designing Memorable Experiences*. Norderstedt: Books on Demand.

Burke, B. (2014) *Gamify: How Gamification Motivates People to do Extraordinary Things*. Brookline, MA: Bibliomotion.

Brás, J.M., Costa, C. and Buhalis, D. (2013) Network analysis and wine routes: The case of the Bairrada Wine Route. In N. Scott and E. Law (eds) *Advances in Service Network Analysis* (pp. 47–68). New York: Routledge.

Dunman, T. and Mattila, A.S. (2005) The role of affective factors on perceived cruise vacation value. *Tourism Management* 26, 311–323.

Fan, D., Buhalis, D. and Lin, B. (2019) A tourist typology of online and face-to-face social contact: Destination immersion and tourism encapsulation/decapsulation. *Annals of Tourism Research* 78. DOI: 10.1016/j.annals.2019.102757.

Farber, M.E. and Hall, T.E. (2007) Emotion and environment: Visitors extraordinary experiences along the Dalton highway in Alaska. *Journal of Leisure Research* 39 (2), 248–270.

Holbrook, M.B. and Hirschman, E.C. (1982) The experiential aspects of consumption: Fantasies, feelings and fun. *Journal of Consumer Research* 9 (2), 132–139.

Huotari, K. and Hamari, J. (2012) Defining gamification: A service marketing perspective. Proceedings of the 16th International Academic MindTrek Conference, 3–5 October, Tampere, Finland: ACM.
Kim, H., Ritchie, J.R. and Mccormick, B. (2012) Development of a scale to measure memorable tourism experience. *Journal of Travel Research* 51 (1), 12–25.
Matthew C. (2020) How artifical intellegience is changing the gaming industry. Available at: https://unbabel.com/blog/author/matthew-carrozo/ (accessed 24 December 2020).
Mistilis, N., Buhalis, D. and Gretzel, U. (2014) eDestination Marketing of the future: The perspective of an Australian tourism stakeholder network. *Journal Travel Research* 53, 1–13.
Nicholson, S. (2012) A user-centered theoretical framework for meaningful gamification. *Games+Learning+Society 8.0*. Madison, USA.
Pine, B.J. and Gilmore, J.H. (1999) *The Experience Economy: Work is Theatre and Every Business is a Stage*. Boston MA: HBS Press.
Robertson, M. (2010) Can't play, won't play. Hide&Seek. Available at: http://hideandseek.net/2010/10/06/cant-play-wont-play/S (accessed 10 October 2014).
Ryan, R.M. and Deci, E.L. (2000) Intrinsic and extrinsic motivations: Classic definitions and new directions. *Contemporary Educational Psychology* 25, 54–67.
Seaborn, K. and Fels, D.I. (2015) Gamification in thoery and action: A survey. *International Journal of Human-Computer Studies* 74, 14–31.
Tung, V.W.S. and Ritchie, J.R.B. (2011) Exploring the essence of memorable tourism experiences. *Annals of Tourism Research* 38 (4), 1367–1386.
Uriely, N. (2005) The tourist expereince: Conceptual developments. *Annals of Tourism Research* 31 (1), 199–216.
Werbach, K. and Hunter, D. (2012) *For the Win: How Game Thinking Can Revolutionize Your Business*. Philadelphia, PA: Wharton Digital Press.
Xu, F., Tian, F., Buhalis, D., Weber, J. and Zhang, H. (2016) Tourists as mobile gamers, the gamification for tourism marketing. *Journal of Travel and Tourism Marketing* 33 (8), 1124–1142.
Xu, F., Buhalis, D. and Weber, J. (2017) Serious games and the gamification of tourism. *Tourism Management* 60, 244–256.
Yovcheva, Z., Buhalis, D., Gatzidis, C. and van Elzakker, C. (2014) Empirical evaluation of smartphone augmented reality browsers in an urban tourism destination context. *International Journal of Mobile Human Computer Interaction* 6 (2), 10–31.

Index

For Product Safety Concerns and Information please contact our EU Authorised Representative:

Easy Access System Europe

Mustamäe tee 50

10621 Tallinn

Estonia

gpsr.requests@easproject.com

www.ingramcontent.com/pod-product-compliance
Ingram Content Group UK Ltd.
Pitfield, Milton Keynes, MK11 3LW, UK
UKHW021827170926
8656IPUK00011B/9

9781845418212